Learn, Practice, Succeed

Eureka Math®
Grade 6
Module 6

Published by Great Minds®.

Copyright © 2019 Great Minds®.

Printed in the U.S.A.

This book may be purchased from the publisher at eureka-math.org.

5 6 7 8 9 10 LSC 26 25 24 23 22 21

ISBN 978-1-64054-969-2

G6-M6-LPS-05.2019

Students, families, and educators:

Thank you for being part of the *Eureka Math®* community, where we celebrate the joy, wonder, and thrill of mathematics.

In *Eureka Math* classrooms, learning is activated through rich experiences and dialogue. That new knowledge is best retained when it is reinforced with intentional practice. The *Learn, Practice, Succeed* book puts in students' hands the problem sets and fluency exercises they need to express and consolidate their classroom learning and master grade-level mathematics. Once students learn and practice, they know they can succeed.

What is in the Learn, Practice, Succeed book?

Fluency Practice: Our printed fluency activities utilize the format we call a Sprint. Instead of rote recall, Sprints use patterns across a sequence of problems to engage students in reasoning and to reinforce number sense while building speed and accuracy. Sprints are inherently differentiated, with problems building from simple to complex. The tempo of the Sprint provides a low-stakes adrenaline boost that increases memory and automaticity.

Classwork: A carefully sequenced set of examples, exercises, and reflection questions support students' in-class experiences and dialogue. Having classwork preprinted makes efficient use of class time and provides a written record that students can refer to later.

Exit Tickets: Students show teachers what they know through their work on the daily Exit Ticket. This check for understanding provides teachers with valuable real-time evidence of the efficacy of that day's instruction, giving critical insight into where to focus next.

Homework Helpers and Problem Sets: The daily Problem Set gives students additional and varied practice and can be used as differentiated practice or homework. A set of worked examples, Homework Helpers, support students' work on the Problem Set by illustrating the modeling and reasoning the curriculum uses to build understanding of the concepts the lesson addresses.

Homework Helpers and Problem Sets from prior grades or modules can be leveraged to build foundational skills. When coupled with *Affirm®*, *Eureka Math's* digital assessment system, these Problem Sets enable educators to give targeted practice and to assess student progress. Alignment with the mathematical models and language used across *Eureka Math* ensures that students notice the connections and relevance to their daily instruction, whether they are working on foundational skills or getting extra practice on the current topic.

Where can I learn more about Eureka Math *resources?*

The Great Minds® team is committed to supporting students, families, and educators with an ever-growing library of resources, available at eureka-math.org. The website also offers inspiring stories of success in the *Eureka Math* community. Share your insights and accomplishments with fellow users by becoming a *Eureka Math* Champion.

Best wishes for a year filled with "aha" moments!

Jill Diniz

Jill Diniz
Chief Academic Officer, Mathematics
Great Minds

Contents

Module 6: Statistics

Example 1: Using Data to Answer Questions

Honeybees are important because they produce honey and pollinate plants. Since 2007, there has been a decline in the honeybee population in the United States. Honeybees live in hives, and a beekeeper in Wisconsin notices that this year, he has 5 fewer hives of bees than last year. He wonders if other beekeepers in Wisconsin are also losing hives. He decides to survey other beekeepers and ask them if they have fewer hives this year than last year, and if so, how many fewer. He then uses the data to conclude that most beekeepers have fewer hives this year than last and that a typical decrease is about 4 hives.

Statistics is about using data to answer questions. In this module, you will use the following four steps in your work with data:

Step 1: Pose a question that can be answered by data.

Step 2: Determine a plan to collect the data.

Step 3: Summarize the data with graphs and numerical summaries.

Step 4: Answer the question posed in Step 1 using the data and summaries.

You will be guided through this process as you study these lessons. This first lesson is about the first step: What is a statistical question, and what does it mean that a question can be answered by data?

Example 2: What Is a Statistical Question?

Jerome, a sixth grader at Roosevelt Middle School, is a huge baseball fan. He loves to collect baseball cards. He has cards of current players and of players from past baseball seasons. With his teacher's permission, Jerome brought his baseball card collection to school. Each card has a picture of a current or past major league baseball player, along with information about the player. When he placed his cards out for the other students to see, they asked Jerome all sorts of questions about his cards. Some asked:

- What is Jerome's favorite card?
- What is the typical cost of a card in Jerome's collection? For example, what is the average cost of a card?
- Are more of Jerome's cards for current players or for past players?
- Which card is the newest card in Jerome's collection?

Lesson 1:   Posing Statistical Questions

1

Exercises 1–5

1. For each of the following, determine whether or not the question is a statistical question. Give a reason for your answer.

   a. Who is my favorite movie star?

   b. What are the favorite colors of sixth graders in my school?

   c. How many years have students in my school's band or orchestra played an instrument?

   d. What is the favorite subject of sixth graders at my school?

   e. How many brothers and sisters does my best friend have?

2. Explain why each of the following questions is not a statistical question.

   a. How old am I?

   b. What's my favorite color?

   c. How old is the principal at our school?

Lesson 1:   Posing Statistical Questions

EUREKA
MATH

3. Ronnie, a sixth grader, wanted to find out if he lived the farthest from school. Write a statistical question that would help Ronnie find the answer.

4. Write a statistical question that can be answered by collecting data from students in your class.

5. Change the following question to make it a statistical question: How old is my math teacher?

Example 3: Types of Data

We use two types of data to answer statistical questions: numerical data and categorical data. If you recorded the ages of 25 baseball cards, we would have numerical data. Each value in a numerical data set is a number. If we recorded the team of the featured player for each of 25 baseball cards, you would have categorical data. Although you still have 25 data values, the data values are not numbers. They would be team names, which you can think of as categories.

Exercises 6–7

6. Identify each of the following data sets as categorical (C) or numerical (N).

    a. Heights of 20 sixth graders _____

    b. Favorite flavor of ice cream for each of 10 sixth graders _____

    c. Hours of sleep on a school night for each of 30 sixth graders _____

    d. Type of beverage drunk at lunch for each of 15 sixth graders _____

    e. Eye color for each of 30 sixth graders _____

    f. Number of pencils in the desk of each of 15 sixth graders _____

7. For each of the following statistical questions, identify whether the data Jerome would collect to answer the question would be numerical or categorical. Explain your answer, and list four possible data values.

    a. How old are the cards in the collection?

    b. How much did the cards in the collection cost?

    c. Where did Jerome get the cards in the collection?

Lesson Summary

Statistics is about using data to answer questions. In this module, the following four steps summarize your work with data:

        Step 1: Pose a question that can be answered by data.

        Step 2: Determine a plan to collect the data.

        Step 3: Summarize the data with graphs and numerical summaries.

        Step 4: Answer the question posed in Step 1 using the data and summaries.

A statistical question is one that can be answered by collecting data and where there will be variability in the data.

Two types of data are used to answer statistical questions: numerical and categorical.

Name _____   Date _____

1. Indicate whether each of the following two questions is a statistical question. Explain why or why not.

   a. How much does Susan's dog weigh?

   b. How much do the dogs belonging to students at our school weigh?

2. If you collected data on the weights of dogs, would the data be numerical or categorical? Explain how you know the data are numerical or categorical.

1. For each of the following, determine whether the question is a statistical question. Give a reason for your answer.

    a. How many bricks are in this wall?

    This is not a statistical question because this question is not answered by collecting data that vary.

    > A *statistical question* is one that can be answered by collecting data, and it is anticipated that the data (information) collected to answer the question will vary.

    > To answer this question, I can just count the bricks. I don't have to collect data since there is just one answer.

    b. On average, how old are the dogs that live on this street?

    This is a statistical question because it would be answered by collecting data on the ages of all the dogs, and there is variability in the ages of the dogs.

    > I anticipate variability in the data because the dogs on the street are likely a variety of ages (e.g., some dogs are young; some dogs are old).

    c. How many days are there until summer break?

    This is not a statistical question because this question is not answered by collecting data that vary.

    d. How many minutes do sixth graders typically spend outside every week?

    This is a statistical question because it would be answered by collecting data on the number of minutes sixth graders spend outside every week, and we expect variability in how many minutes are recorded for each student. They will not all be the same.

2. Identify each of the following data sets as categorical (C) or numerical (N). Explain your answer.

    a. Height of sixth graders

    N; the height can be measured as number of inches, for example, so the data set is numerical.

    > In a numerical data set, each value is a number.

    b. The hair color of 20 adults

    C; hair color is categorical because hair colors are categories.

    > In a categorical data set, the values are categories, not numbers.

3. Rewrite the following question as a statistical question:
    How many minutes do you spend on homework each week?

    Answers may vary. A sample response is provided below.

    How many minutes do sixth grade students typically spend on homework each week?

4. Write a statistical question that would be answered by collecting data from the animals that reside at the Bambelela Wildlife Sanctuary?

    Answers may vary. A sample response is provided below.

    How long do the cheetahs spend at the watering hole in July?

    > Because different cheetahs will spend different amounts of time at the watering hole in July, this is a statistical question because the data that will be collected will vary.

5. Are the data you would collect to answer the question you wrote in Problem 4 categorical or numerical? Explain your answer.

    Numerical. The time at the watering hole can be measured in minutes, for example, so each value in the data set is a number, and the data set is numerical. Note: The answer to Problem 5 depends on the statistical question asked in Problem 4.

EUREKA MATH

1. For each of the following, determine whether the question is a statistical question. Give a reason for your answer.
    a. How many letters are in my last name?
    b. How many letters are in the last names of the students in my sixth-grade class?
    c. What are the colors of the shoes worn by students in my school?
    d. What is the maximum number of feet that roller coasters drop during a ride?
    e. What are the heart rates of students in a sixth-grade class?
    f. How many hours of sleep per night do sixth graders usually get when they have school the next day?
    g. How many miles per gallon do compact cars get?

2. Identify each of the following data sets as categorical (C) or numerical (N). Explain your answer.
    a. Arm spans of 12 sixth graders
    b. Number of languages spoken by each of 20 adults
    c. Favorite sport of each person in a group of 20 adults
    d. Number of pets for each of 40 third graders
    e. Number of hours a week spent reading a book for a group of middle school students

3. Rewrite each of the following questions as a statistical question.
    a. How many pets does your teacher have?
    b. How many points did the high school soccer team score in its last game?
    c. How many pages are in our math book?
    d. Can I do a handstand?

4. Write a statistical question that would be answered by collecting data from the sixth graders in your classroom.

5. Are the data you would collect to answer the question you wrote in Problem 4 categorical or numerical? Explain your answer.

Lesson 1:   Posing Statistical Questions

11

Example 1: Heart Rate

Mia, a sixth grader at Roosevelt Middle School, was thinking about joining the middle school track team. She read that Olympic athletes have lower resting heart rates than most people. She wondered about her own heart rate and how it would compare to other students. Mia was interested in investigating the statistical question: What are the heart rates of students in my sixth-grade class?

Heart rates are expressed as beats per minute (or bpm). Mia knew her resting heart rate was 80 beats per minute. She asked her teacher if she could collect the heart rates of the other students in her class. With the teacher's help, the other sixth graders in her class found their heart rates and reported them to Mia. The following numbers are the resting heart rates (in beats per minute) for the 22 other students in Mia's class.

89 87 85 84 90 79 83 85 86 88 84 81 88 85 83 83 86 82 83 86 82 84

Exercises 1–10

1. What was the heart rate for the student with the lowest heart rate?

2. What was the heart rate for the student with the highest heart rate?

3. How many students had a heart rate greater than 86 bpm?

4. What fraction of students had a heart rate less than 82 bpm?

5. What heart rate occurred most often?

6. What heart rate describes the center of the data?

7. Some students had heart rates that were unusual in that they were quite a bit higher or quite a bit lower than most other students' heart rates. What heart rates would you consider unusual?

8. If Mia's teacher asked what the typical heart rate is for sixth graders in the class, what would you tell Mia's teacher?

9. Remember that Mia's heart rate was 80 bpm. Add a dot for Mia's heart rate to the dot plot in Example 1.

10. How does Mia's heart rate compare with the heart rates of the other students in the class?

EUREKA
MATH

Example 2: Seeing the Spread in Dot Plots

Mia's class collected data to answer several other questions about her class. After collecting the data, they drew dot plots of their findings.

One student collected data to answer the question: How many textbooks are in the desks or lockers of sixth graders? She made the following dot plot, not including her data.

Dot Plot of Number of Textbooks

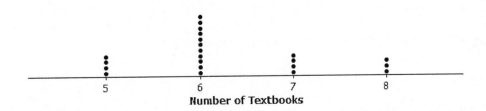

Number of Textbooks

Another student in Mia's class wanted to ask the question: How tall are the sixth graders in our class?

This dot plot shows the heights of the sixth graders in Mia's class, not including the datum for the student conducting the survey.

Dot Plot of Height

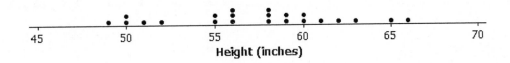

Height (inches)

Exercises 11–14

Below are four statistical questions and four different dot plots of data collected to answer these questions. Match each statistical question with the appropriate dot plot, and explain each choice.

Statistical Questions:

11. What are the ages of fourth graders in our school?

12. What are the heights of the players on the eighth-grade boys' basketball team?

13. How many hours of TV do sixth graders in our class watch on a school night?

14. How many different languages do students in our class speak?

Dot Plot A   **Dot Plot B**

Dot Plot C   **Dot Plot D**

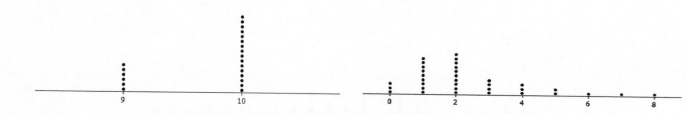

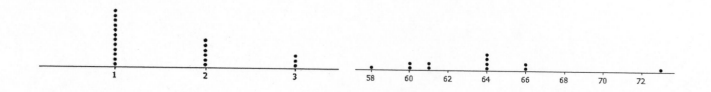

Lesson 2:   Displaying a Data Distribution

EUREKA
MATH

Name _____   Date _____

A sixth-grade class collected data on the number of letters in the first names (name lengths) of all the students in class. Here is the dot plot of the data they collected:

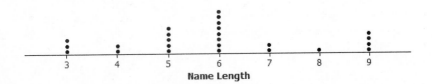

Name Length

1. How many students are in the class?

2. What is the shortest name length?

3. What is the longest name length?

4. What name length occurs most often?

5. What name length describes the center of the data?

1. The dot plot below shows the number of siblings of the sixth grade students in Ms. Baker's class.

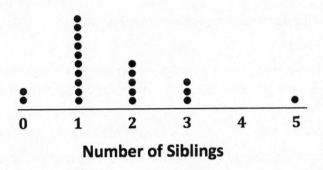

Number of Siblings

a. What statistical question do you think could be answered using these data?

    How many siblings does the typical sixth grader have?

    To answer this question, I have to collect data, and I anticipate there is variability in the data set since every sixth grader will not have the same number of siblings.

b. What was the most number of siblings recorded in the class?

    5 *siblings*

c. What was the least number of siblings recorded in the class?

    0 *siblings*

d. What is the most common number of siblings (the number of siblings that occurred most often)?

    1 *sibling*

    I can see that the most common number of siblings students recorded is 1 because the number of dots for that number is more than any other number.

e. How many students recorded the most common number of siblings?

    10 *students*

f. How many students had more than 2 siblings?

    4 *students*

    I can count the number of dots for 3, 4, and 5 siblings.

g. If a new student joins the class and has 1 sibling, how does this student compare with the other students?

   The new student would have the most common number of siblings.

2. Read each of the following statistical questions. Write a description of what the dot plot of the data collected to answer the question might look like. Your description should include a description of the spread of the data and the center of the data.

   a. How many minutes do sixth graders spend in the cafeteria eating lunch during a typical school day?

   Most sixth graders are in the cafeteria for the same number of minutes, so the spread would be small. Differences may exist for those students who have to leave early from school and do not eat lunch in the cafeteria. Student responses vary based on their estimates of the number of minutes sixth graders spend in the cafeteria eating lunch during the school day.

   b. What is the number of books owned by the sixth graders in our class?

   These data would have a very big spread. Some students might have very few books, while others could have many books. A typical value of 10 (or something similar) would identify a center. In this case, the center is based on the number most commonly reported by students.

   > For the center, students may describe a number that occurs most often, the number in the middle, or the average. It is important to gauge students' thinking about what center means.

EUREKA
MATH

1. The dot plot below shows the vertical jump height (in inches) of some NBA players. A vertical jump height is how high a player can jump from a standstill.

Dot Plot of Vertical Jump

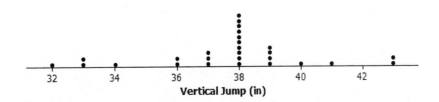

Vertical Jump (in)

   a. What statistical question do you think could be answered using these data?

   b. What was the highest vertical jump by a player?

   c. What was the lowest vertical jump by a player?

   d. What is the most common vertical jump height (the height that occurred most often)?

   e. How many players jumped the most common vertical jump height?

   f. How many players jumped higher than 40 inches?

   g. Another NBA player jumped 33 inches. Add a dot for this player on the dot plot. How does this player compare with the other players?

2. Below are two statistical questions and two different dot plots of data collected to answer these questions. Match each statistical question with its dot plot, and explain each choice.

   Statistical Questions:

   a. What is the number of fish (if any) that students in class have in an aquarium at their homes?

   b. How many days out of the week do the children on my street go to the playground?

Dot Plot A	Dot Plot B

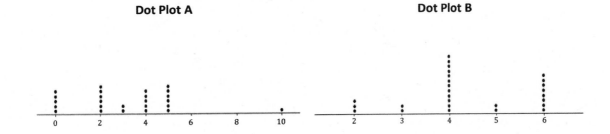

3. Read each of the following statistical questions. Write a description of what the dot plot of data collected to answer the question might look like. Your description should include a description of the spread of the data and the center of the data.

    a. What is the number of hours sixth graders are in school during a typical school day?

    b. What is the number of video games owned by the sixth graders in our class?

EUREKA
MATH

Example 1: Hours of Sleep

Robert, a sixth grader at Roosevelt Middle School, usually goes to bed around 10:00 p.m. and gets up around 6:00 a.m. to get ready for school. That means he gets about 8 hours of sleep on a school night. He decided to investigate the statistical question: How many hours per night do sixth graders usually sleep when they have school the next day?

Robert took a survey of 29 sixth graders and collected the following data to answer the question.

7 8 5 9 9 9 7 7 10 10 11 9 8 8 8 12 6 11 10 8 8 9 9 9 8 10 9 9 8

Robert decided to make a dot plot of the data to help him answer his statistical question. Robert first drew a number line and labeled it from 5 to 12 to match the lowest and highest number of hours slept. Robert's datum is not included.

Dot Plot of Number of Hours Slept

5	6	7	8	9	10	11	12

Number of Hours Slept

He then placed a dot above 7 for the first value in the data set. He continued to place dots above the numbers until each number in the data set was represented by a dot.

Dot Plot of Number of Hours Slept

5	6	7	8	9	10	11	12

Number of Hours Slept

Exercises 1–9

1. Complete Robert's dot plot by placing a dot above the corresponding number on the number line for each value in the data set. If there is already a dot above a number, then add another dot above the dot already there. Robert's datum is not included.

2. What are the least and the most hours of sleep reported in the survey of sixth graders?

3. What number of hours slept occurred most often in the data set?

4. What number of hours of sleep would you use to describe the center of the data?

5. Think about how many hours of sleep you usually get on a school night. How does your number compare with the number of hours of sleep from the survey of sixth graders?

Here are the data for the number of hours the sixth graders usually sleep when they do not have school the next day.

7 8 10 11 5 6 12 13 13 7 9 8 10 12 11 12 8 9 10 11 10 12 11 11 11 12 11 11 10

6. Make a dot plot of the number of hours slept when there is no school the next day.

7. When there is no school the next day, what number of hours of sleep would you use to describe the center of the data?

8. What are the least and most number of hours slept with no school the next day reported in the survey?

EUREKA MATH

9. Do students tend to sleep longer when they do not have school the next day than when they do have school the next day? Explain your answer using the data in both dot plots.

Example 2: Building and Interpreting a Frequency Table

A group of sixth graders investigated the statistical question, "How many hours per week do sixth graders spend playing a sport or an outdoor game?"

Here are the data students collected from a sample of 26 sixth graders showing the number of hours per week spent playing a sport or a game outdoors.

3 2 0 6 3 3 3 1 1 2 2 8 12 4 4 4 3 3 1 1 0 0 6 2 3 2

To help organize the data, students summarized the data in a frequency table. A frequency table lists possible data values and how often each value occurs.

To build a frequency table, first make three columns. Label one column "Number of Hours Playing a Sport/Game," label the second column "Tally," and label the third column "Frequency." Since the least number of hours was 0 and the most was 12, list the numbers from 0 to 12 in the "Number of Hours" column.

Exercises 10–15

10. Complete the tally mark column in the table created in Example 2.

11. For each number of hours, find the total number of tally marks, and place this in the frequency column in the table created in Example 2.

12. Make a dot plot of the number of hours playing a sport or playing outdoors.

13. What number of hours describes the center of the data?

14. How many of the sixth graders reported that they spend eight or more hours a week playing a sport or playing outdoors?

15. The sixth graders wanted to answer the question, "How many hours do sixth graders spend per week playing a sport or playing an outdoor game?" Using the frequency table and the dot plot, how would you answer the sixth graders' question?

Name _____   Date _____

A biologist collected data to answer the question, "How many eggs do robins lay?"

The following is a frequency table of the data she collected:

Number of Eggs	Tally	Frequency
1	||	
2	ⵜ⵿ⵜ ⵜ⵿ⵜ ||	
3	ⵜ⵿ⵜ ⵜ⵿ⵜ ⵜ⵿ⵜ |||	
4	ⵜ⵿ⵜ ||||	
5	|	

1. Complete the frequency column.

2. Draw a dot plot of the data on the number of eggs a robin lays.

3. What number of eggs describes the center of the data?

1. The data below are the number of goals scored by a professional indoor soccer team over its last 21 games.

   6 7 7 8 8 8 8 8 8 9 9 9 10 10 10 10 11 11 11 12 14

   a. Make a dot plot of the number of goals scored.

   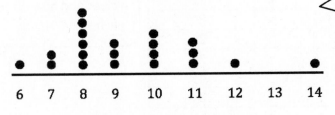

   Dot Plot of Number of Soccer Goals

   Number of Soccer Goals Scored

   > The data values are already in order from least to greatest. Since the smallest number is 6 and the largest number is 14, I can list the numbers on the scale sequentially, starting at 6, ending at 14, and counting by 1. I can use the given data to plot the points.

   b. What number of goals describes the center of the data?

   The center of the data is around 9. (Answers may vary, but student responses should be around the center of the data distribution.)

   > Since there are 21 values in the data set, the center of the data will be the 11th value, which is 9.

   c. What is the least and most number of goals scored by the team?

   The least number of goals scored is 6, and 14 is the most.

   d. Over the 21 games played, the team lost 9 games. Circle the dots on the plot that you think represent the games that the team lost. Explain your answer.

   Students will most likely circle the lowest 9 scores, but answers may vary. Students need to supply an explanation in order to defend their answers.

2. A sixth grader collected data on the number of hours 15 students read independently each week. The following are the number of hours for the 15 students:

3 2 4 4 5 7 6 3 6 3 7 6 6 7 1

a. Complete the frequency table.

Number of Hours	Tally	Frequency
1	|	1
2	|	1
3	| | |	3
4	| |	2
5	|	1
6	| | | |	4
7	| | |	3

A frequency table lists possible data values and how often each value occurs.

b. What number of hours describes the center of the data?

5

c. What number of hours occurs most often for these 15 students?

6

Lesson 3:   Creating a Dot Plot

EUREKA MATH

1. The data below are the number of goals scored by a professional indoor soccer team over its last 23 games.

    8 16 10 9 11 11 10 15 16 11 15 13 8 9 11 9 8 11 16 15 10 9 12

    a. Make a dot plot of the number of goals scored.

    b. What number of goals describes the center of the data?

    c. What is the least and most number of goals scored by the team?

    d. Over the 23 games played, the team lost 10 games. Circle the dots on the plot that you think represent the games that the team lost. Explain your answer.

2. A sixth grader rolled two number cubes 21 times. The student found the sum of the two numbers that he rolled each time. The following are the sums for the 21 rolls of the two number cubes.

    9 2 4 6 5 7 8 11 9 4 6 5 7 7 8 8 7 5 7 6 6

    a. Complete the frequency table.

Sum Rolled	Tally	Frequency
2		
3		
4		
5		
6		
7		
8		
9		
10		
11		
12		

    b. What sum describes the center of the data?

    c. What sum occurred most often for these 21 rolls of the number cubes?

3. The dot plot below shows the number of raisins in 25 small boxes of raisins.

Dot Plot of Number of Raisins

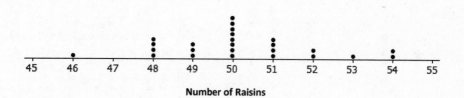

Number of Raisins

a. Complete the frequency table.

Number of Raisins	Tally	Frequency
46		
47		
48		
49		
50		
51		
52		
53		
54		

b. Another student opened up a box of raisins and reported that it had 63 raisins. Do you think that this student had the same size box of raisins? Why or why not?

EUREKA MATH

Example 1: Frequency Table with Intervals

The boys' and girls' basketball teams at Roosevelt Middle School wanted to raise money to help buy new uniforms. They decided to sell baseball caps with the school logo on the front to family members and other interested fans. To obtain the correct cap size, students had to measure the head circumference (distance around the head) of the adults who wanted to order a cap. The following data set represents the head circumferences, in millimeters (mm), of the adults.

513, 525, 531, 533, 535, 535, 542, 543, 546, 549, 551, 552, 552, 553, 554, 555, 560, 561, 563, 563, 563, 565, 565, 568, 568, 571, 571, 574, 577, 580, 583, 583, 584, 585, 591, 595, 598, 603, 612, 618

The caps come in six sizes: XS, S, M, L, XL, and XXL. Each cap size covers an interval of head circumferences. The cap manufacturer gave students the table below that shows the interval of head circumferences for each cap size. The interval 510–< 530 represents head circumferences from 510 mm to 530 mm, not including 530.

Cap Sizes	Interval of Head Circumferences (millimeters)	Tally	Frequency
XS	510–< 530		
S	530–< 550		
M	550–< 570		
L	570–< 590		
XL	590–< 610		
XXL	610–< 630		

Exercises 1–4

1. What size cap would someone with a head circumference of 570 mm need?

2. Complete the tally and frequency columns in the table in Example 1 to determine the number of each size cap students need to order for the adults who wanted to order a cap.

3. What head circumference would you use to describe the center of the data?

4. Describe any patterns that you observe in the frequency column.

Example 2: Histogram

One student looked at the tally column and said that it looked somewhat like a bar graph turned on its side. A histogram is a graph that is like a bar graph except that the horizontal axis is a number line that is marked off in equal intervals.

To make a histogram:

- Draw a horizontal line, and mark the intervals.
- Draw a vertical line, and label it Frequency.
- Mark the Frequency axis with a scale that starts at 0 and goes up to something that is greater than the largest frequency in the frequency table.
- For each interval, draw a bar over that interval that has a height equal to the frequency for that interval.

The first two bars of the histogram have been drawn below.

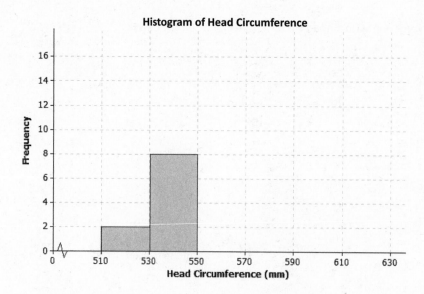

© 2019 Great Minds®. eureka-math.org

Exercises 5–9

5. Complete the histogram by drawing bars whose heights are the frequencies for the other intervals.

6. Based on the histogram, describe the center of the head circumferences.

7. How would the histogram change if you added head circumferences of 551 mm and 569 mm to the data set?

8. Because the 40 head circumference values were given, you could have constructed a dot plot to display the head circumference data. What information is lost when a histogram is used to represent a data distribution instead of a dot plot?

9. Suppose that there had been 200 head circumference measurements in the data set. Explain why you might prefer to summarize this data set using a histogram rather than a dot plot.

Example 3: Shape of the Histogram

A histogram is useful to describe the shape of the data distribution. It is important to think about the shape of a data distribution because depending on the shape, there are different ways to describe important features of the distribution, such as center and variability.

A group of students wanted to find out how long a certain brand of AA batteries lasted. The histogram below shows the data distribution for how long (in hours) that some AA batteries lasted. Looking at the shape of the histogram, notice how the data mound up around a center of approximately 105 hours. We would describe this shape as mound shaped or symmetric. If we were to draw a line down the center, notice how each side of the histogram is approximately the same, or a mirror image of the other. This means the histogram is approximately symmetrical.

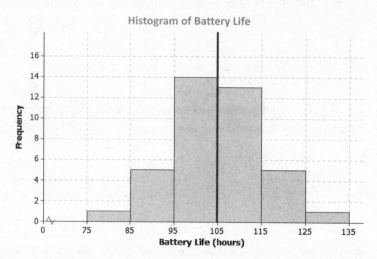

Another group of students wanted to investigate the maximum drop length for roller coasters. The histogram below shows the maximum drop (in feet) of a selected group of roller coasters. This histogram has a skewed shape. Most of the data are in the intervals from 50 feet to 170 feet. But there is one value that falls in the interval from 290 feet to 330 feet and one value that falls in the interval from 410 feet to 550 feet. These two values are unusual (or not typical) when compared to the rest of the data because they are much greater than most of the data.

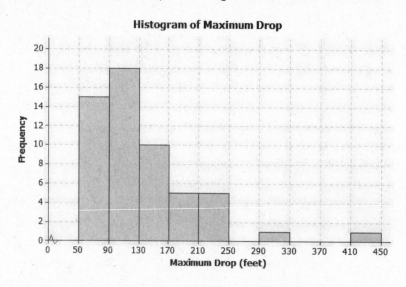

EUREKA MATH

Exercises 10–12

10. The histogram below shows the highway miles per gallon of different compact cars.

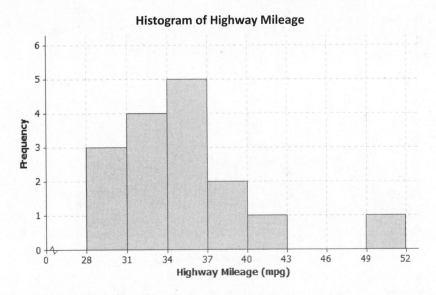

Histogram of Highway Mileage

a. Describe the shape of the histogram as approximately symmetric, skewed left, or skewed right.

b. Draw a vertical line on the histogram to show where the typical number of miles per gallon for a compact car would be.

c. What does the shape of the histogram tell you about miles per gallon for compact cars?

11. Describe the shape of the head circumference histogram that you completed in Exercise 5 as approximately symmetric, skewed left, or skewed right.

12. Another student decided to organize the head circumference data by changing the width of each interval to be 10 instead of 20. Below is the histogram that the student made.

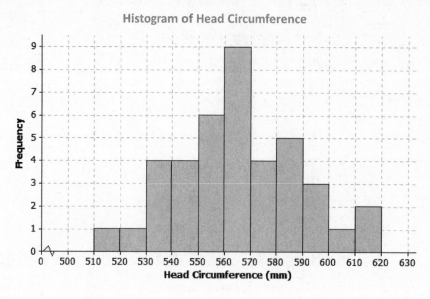

a. How does this histogram compare with the histogram of the head circumferences that you completed in Exercise 5?

b. Describe the shape of this new histogram as approximately symmetric, skewed left, or skewed right.

c. How many head circumferences are in the interval from 570 to 590 mm?

d. In what interval would a head circumference of 571 mm be included? In what interval would a head circumference of 610 mm be included?

EUREKA MATH

Name _____   Date _____

The frequency table below shows the length of selected movies shown in a local theater over the past six months.

Length of Movie (minutes)	Tally	Frequency
80–< 90	|	1
90–< 100	||||	4
100–< 110	卌 ||	7
110–< 120	卌	5
120–< 130	卌 ||	7
130–< 140	|||	3
140–< 150	|	1

1. Construct a histogram for the length of movies data.

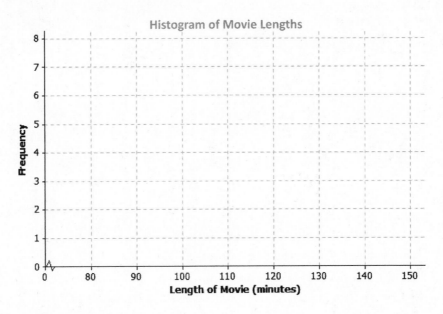

Histogram of Movie Lengths

2. Describe the shape of the histogram.

3. What does the histogram tell you about the length of movies?

1. The following histogram summarizes the ages of the participants in a community relay race.

Ages of Participants in Relay Race

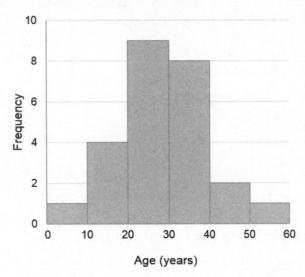

I can look for the interval that has the greatest frequency.

a. Which age interval contains the most participants? How many participants are represented in that interval?

   The interval 20 to 30 contains the most participants. There are 9 participants whose ages fall into that category.

b. Describe the shape of the histogram.

   The shape of the histogram is approximately symmetrical.

c. What does the histogram tell you about the ages of the participants in the relay race?

   Most of the ages are between 20 and 40, with a few people with ages much smaller or larger than the rest.

d. Which interval describes the center of the ages of the participants?

The interval of 20 to 40 describes the center of the ages. Since this data distribution is approximately symmetrical, the center is probably around 30. (Answers may vary, but student responses should be around the center of the data distribution.)

e. An age of 19 would be included in which interval?

An age of 19 is in the interval from 10 to 20.

2. Listed are the prices for various items sold at a garage sale.

$4 $8 $16 $17 $16 $18 $11 $26

$34 $28 $23 $15 $10 $30 $29 $13

a. Complete the frequency table using the given intervals of width 5.

Prices for Items Sold	Tally	Frequency
$0—< $5	I	1
$5—< $10	I	1
$10—< $15	III	3
$15—< $20	HHH	5
$20—< $25	I	1
$25—< $30	III	3
$30—< $35	II	2

For the interval $0—< $5, I can look for the prices, $0, $1, $2, $3, and $4 in the data set. There is only one price, $4, that fits in this category.

EUREKA
MATH®

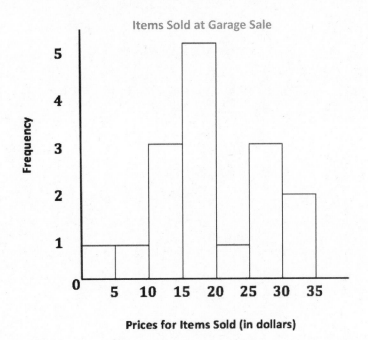

c. Describe the center and shape of the histogram.

The center is around 17; the histogram is mound shaped and skewed slightly to the left.

(Answers may vary, but student responses for describing the center should be around the center of the data distribution.)

d. In the frequency table below, the intervals are changed. Using the garage sale data above, complete the frequency table with intervals of width 10.

Prices for Items Sold	Tally	Frequency
$0–< $10	||	2
$10–< $20	⊬⊬ |||	8
$20–< $30	||||	4
$30–< $40	||	2

Even though the data is the same, I can see how the frequency table looks different because the width of the interval has changed.

e. Draw a histogram.

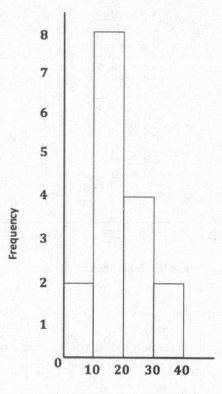

Items Sold at Garage Sale

Frequency (y-axis), Prices for Items Sold (in dollars) (x-axis)

3. Use the histograms that you constructed in Problem 2 parts (b) and (e) to answer the following questions.

a. Why are there fewer bars in the histogram in part (e) than the histogram in part (b)?

There are fewer bars in part (e) because the width of the interval changed from $5 to $10.

b. Did the shape of the histogram in part (e) change from the shape of the histogram in part (b)?

Generally, both are mound shaped, but the histogram in part (e) is skewed less to the left.

c. Did your estimate of the center change from the histogram in part (b) to the histogram in part (e)?

No; the centers of the two histograms are about the same.

EUREKA
MATH

1. The following histogram summarizes the ages of the actresses whose performances have won in the Best Leading Actress category at the annual Academy Awards (i.e., Oscars).

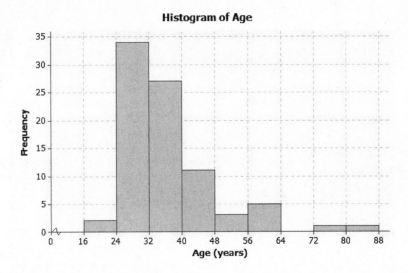

Histogram of Age

   a. Which age interval contains the most actresses? How many actresses are represented in that interval?

   b. Describe the shape of the histogram.

   c. What does the histogram tell you about the ages of actresses who won the Oscar for best actress?

   d. Which interval describes the center of the ages of the actresses?

   e. An age of 72 would be included in which interval?

2. The frequency table below shows the seating capacity of arenas for NBA basketball teams.

Number of Seats	Tally	Frequency
17,000–< 17,500	||	2
17,500–< 18,000	|	1
18,000–< 18,500	⊦⊦⊦⊦ |	6
18,500–< 19,000	⊦⊦⊦⊦	5
19,000–< 19,500	⊦⊦⊦⊦	5
19,500–< 20,000	⊦⊦⊦⊦	5
20,000–< 20,500	||	2
20,500–< 21,000	||	2
21,000–< 21,500		0
21,500–< 22,000		0
22,000–< 22,500	|	1

   a. Draw a histogram for the number of seats in the NBA arenas data. Use the histograms you have seen throughout this lesson to help you in the construction of your histogram.

   b. What is the width of each interval? How do you know?

Lesson 4: Creating a Histogram

45

© 2019 Great Minds®. eureka-math.org

c. Describe the shape of the histogram.

d. Which interval describes the center of the number of seats data?

3. Listed are the grams of carbohydrates in hamburgers at selected fast food restaurants.

| 33 | 40 | 66 | 45 | 28 | 30 | 52 | 40 | 26 | 42 |
| 42 | 44 | 33 | 44 | 45 | 32 | 45 | 45 | 52 | 24 |

a. Complete the frequency table using the given intervals of width 5.

Number of Carbohydrates (grams)	Tally	Frequency
20–< 25		
25–< 30		
30–< 35		
35–< 40		
40–< 45		
45–< 50		
50–< 55		
55–< 60		
60–< 65		
65–< 70		

b. Draw a histogram of the carbohydrate data.

c. Describe the center and shape of the histogram.

d. In the frequency table below, the intervals are changed. Using the carbohydrate data above, complete the frequency table with intervals of width 10.

Number of Carbohydrates (grams)	Tally	Frequency
20–< 30		
30–< 40		
40–< 50		
50–< 60		
60–< 70		

e. Draw a histogram.

4. Use the histograms that you constructed in Exercise 3 parts (b) and (e) to answer the following questions.

a. Why are there fewer bars in the histogram in part (e) than the histogram in part (b)?

b. Did the shape of the histogram in part (e) change from the shape of the histogram in part (b)?

c. Did your estimate of the center change from the histogram in part (b) to the histogram in part (e)?

EUREKA MATH

Example 1: Relative Frequency Table

In Lesson 4, we investigated the head circumferences that the boys' and girls' basketball teams collected. Below is the frequency table of the head circumferences that they measured.

Cap Sizes	Interval of Head Circumferences (millimeters)	Tally	Frequency
XS	510–< 530	||	2
S	530–< 550	⋕ |||	8
M	550–< 570	⋕ ⋕ ⋕	15
L	570–< 590	⋕ ||||	9
XL	590–< 610	||||	4
XXL	610–< 630	||	2
		Total: 40

Isabel, one of the basketball players, indicated that most of the caps were small (S), medium (M), or large (L). To decide if Isabel was correct, the players added a relative frequency column to the table.

Relative frequency is the frequency for an interval divided by the total number of data values. For example, the relative frequency for the extra small (XS) cap is 2 divided by 40, or 0.05. This represents the fraction of the data values that were XS.

Exercises 1–4

1. Complete the relative frequency column in the table below.

Cap Sizes	Interval of Head Circumferences (millimeters)	Tally	Frequency	Relative Frequency
XS	510–< 530	||	2	$\frac{2}{40} = 0.050$
S	530–< 550	卌 |||	8	$\frac{8}{40} = 0.200$
M	550–< 570	卌 卌 卌	15	
L	570–< 590	卌 ||||	9	
XL	590–< 610	||||	4	
XXL	610–< 630	||	2	
			Total: 40	

2. What is the total of the relative frequency column?

3. Which interval has the greatest relative frequency? What is the value?

4. What percentage of the head circumferences are between 530 and 589 mm? Show how you determined the answer.

EUREKA
MATH

Example 2: Relative Frequency Histogram

The players decided to construct a histogram using the relative frequencies instead of the frequencies.

They noticed that the relative frequencies in the table ranged from close to 0 to about 0.40. They drew a number line and marked off the intervals on that line. Then, they drew the vertical line and labeled it Relative Frequency. They added a scale to this line by starting at 0 and counting by 0.05 until they reached 0.40.

They completed the histogram by drawing the bars so the height of each bar matched the relative frequency for that interval. Here is the completed relative frequency histogram:

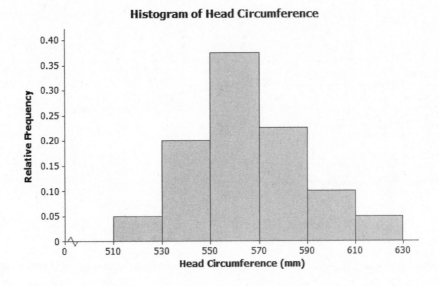

Exercises 5–6

5.

   a. Describe the shape of the relative frequency histogram of head circumferences from Example 2.

   b. How does the shape of this relative frequency histogram compare with the frequency histogram you drew in Exercise 5 of Lesson 4?

   c. Isabel said that most of the caps that needed to be ordered were small (S), medium (M), and large (L). Was she right? What percentage of the caps to be ordered are small, medium, or large?

6. Here is the frequency table of the seating capacity of arenas for the NBA basketball teams.

Number of Seats	Tally	Frequency	Relative Frequency
17,000–< 17,500	||	2	
17,500–< 18,000	|	1	
18,000–< 18,500	⊞⊞ |	6	
18,500–< 19,000	⊞⊞	5	
19,000–< 19,500	⊞⊞	5	
19,500–< 20,000	⊞⊞	5	
20,000–< 20,500	||	2	
20,500–< 21,000	||	2	
21,000–< 21,500		0	
21,500–< 22,000		0	
22,000–< 22,500	|	1	

a. What is the total number of NBA arenas?

b. Complete the relative frequency column. Round the relative frequencies to the nearest thousandth.

c. Construct a relative frequency histogram.

Lesson 5:   Describing a Distribution Displayed in a Histogram

EUREKA
MATH

d. Describe the shape of the relative frequency histogram.

e. What percentage of the arenas have a seating capacity between 18,500 and 19,999 seats?

f. How does this relative frequency histogram compare to the frequency histogram that you drew in Problem 2 of the Problem Set in Lesson 4?

> **Lesson Summary**
>
> A **relative frequency** is the frequency for an interval divided by the total number of data values. For example, if the first interval contains 8 out of a total of 32 data values, the relative frequency of the first interval is $\frac{8}{32} = \frac{1}{4} = 0.25$, or 25%.
>
> A **relative frequency histogram** is a histogram that is constructed using relative frequencies instead of frequencies.

Name _____   Date _____

Calculators are allowed for completing your problems.

Hector's mom had a rummage sale, and after she sold an item, she tallied the amount of money she received for the item. The following is the frequency table Hector's mom created.

Amount of Money Received for the Item	Tally	Frequency	Relative Frequency
$0–< $5	||	2	
$5–< $10	|	1	
$10–< $15	||||	4	
$15–< $20	⊬⊬⊬ ⊬⊬⊬	10	
$20–< $25	⊬⊬⊬	5	
$25–< $30	|||	3	
$30–< $35	||	2	

a. What was the total number of items sold at the rummage sale?

b. Complete the relative frequency column. Round the relative frequencies to the nearest thousandth.

c. What percentage of the items Hector's mom sold were sold for $15 or more but less than $20?

1. Below is a relative frequency histogram of the test scores (in percentage) of a selected group of sixth graders.

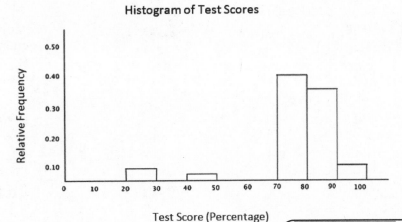

Histogram of Test Scores

a. Describe the shape of the relative frequency histogram.

The shape is skewed to the left.

> This graph is skewed left because it has a tail that is longer on the left side. The graph is skewed toward the smaller values.

b. What does the shape tell you about the maximum test score (in percentage) of the selected group of sixth graders?

The shape tells us most of the sixth graders have a test score that is between 70% and 90% but that some students have a test score that is quite a bit lower than the others.

c. Clara said that more than half of the data values are in the interval from 80% to 100%. Do you agree with Clara? Why or why not?

I do not agree because that interval contains 45% of the data.

> In the interval 80–90, the relative frequency is 0.35 (or 35%). In the interval 90–100, the relative frequency is 0.10 (or 10%). The cumulative relative frequency is $0.35 + 0.10 = 0.45$, which is 45% and less than half.

2. The frequency table below shows the length of selected professional football games over the past 6 months.

Length of Game (minutes)	Tally	Frequency	Relative Frequency
160–< 170	||	2	$\dfrac{2}{25} = 0.08$
170–< 180	|||	3	$\dfrac{3}{25} = 0.12$
180–< 190	⊬⊬ |	6	$\dfrac{6}{25} = 0.24$
190–< 200	||||	4	$\dfrac{4}{25} = 0.16$
200–< 210	⊬⊬ |	6	$\dfrac{6}{25} = 0.24$
210–< 220	|||	3	$\dfrac{3}{25} = 0.12$
220–< 230	|	1	$\dfrac{1}{25} = 0.04$

a. Complete the relative frequency column. Round the relative frequencies to the nearest hundredth.

   See the table above.

   > Relative frequency is the frequency for an interval divided by the total number of data values.

b. What percentage of the football games are greater than or equal to 210 minutes?

$$0.12 + 0.04 = 0.16$$

   16% of the game lengths are greater than or equal to 210 minutes.

   > I can add the relative frequencies for the game lengths greater than or equal to 210 minutes (there are two intervals) and then determine the percentage of games in that category.

EUREKA MATH

c. Draw a relative frequency histogram. (Hint: Label the relative frequency scale starting at 0 and going up to 0.30, marking off intervals of 0.05).

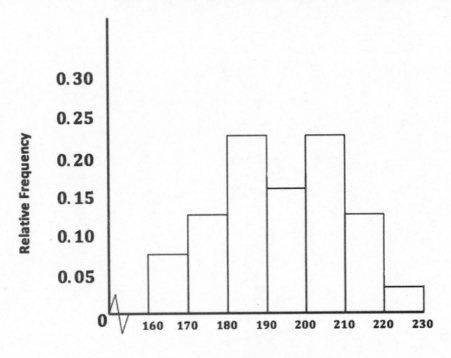

Relative Frequency Histogram of Game Lengths

Relative Frequency

Length of Games (in minutes)

d. Describe the shape of the relative frequency histogram.

   The histogram is mound shaped and approximately symmetric.

e. What does the shape tell you about the length of football games?

   The shape tells us the length of most football games is between 180 and 210 minutes.

EUREKA MATH®

© 2019 Great Minds®. eureka-math.org

1. Below is a relative frequency histogram of the maximum drop (in feet) of a selected group of roller coasters.

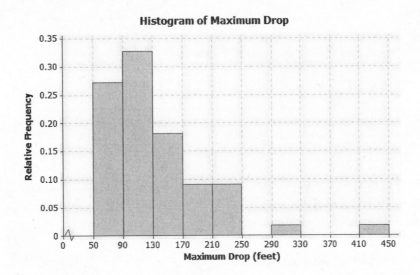

a. Describe the shape of the relative frequency histogram.

b. What does the shape tell you about the maximum drop (in feet) of roller coasters?

c. Jerome said that more than half of the data values are in the interval from 50 to 130 feet. Do you agree with Jerome? Why or why not?

2. The frequency table below shows the length of selected movies shown in a local theater over the past 6 months.

Length of Movie (minutes)	Tally	Frequency	Relative Frequency
80–< 90	|	1	
90–< 100	||||	4	
100–< 110	⦀⦀ ||	7	
110–< 120	⦀⦀	5	
120–< 130	⦀⦀ ||	7	
130–< 140	|||	3	
140–< 150	|	1	

a. Complete the relative frequency column. Round the relative frequencies to the nearest thousandth.

b. What percentage of the movie lengths are greater than or equal to 130 minutes?

c. Draw a relative frequency histogram. (Hint: Label the relative frequency scale starting at 0 and going up to 0.30, marking off intervals of 0.05.)

d. Describe the shape of the relative frequency histogram.

e. What does the shape tell you about the length of movie times?

3. The table below shows the highway miles per gallon of different compact cars.

Mileage	Tally	Frequency	Relative Frequency
28–< 31	|||	3	
31–< 34	||||	4	
34–< 37	⊬⊬	5	
37–< 40	||	2	
40–< 43	|	1	
43–< 46		0	
46–< 49		0	
49–< 52	|	1	

a. What is the total number of compact cars?

b. Complete the relative frequency column. Round the relative frequencies to the nearest thousandth.

c. What percent of the cars get between 31 and up to but not including 37 miles per gallon on the highway?

d. Juan drew the relative frequency histogram of the highway miles per gallon for the compact cars, shown on the right. Did Juan draw the histogram correctly? Explain your answer.

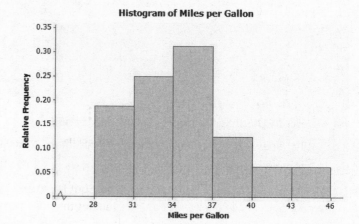

Histogram of Miles per Gallon

Lesson 5:   Describing a Distribution Displayed in a Histogram

EUREKA MATH

Example 1

Recall that in Lesson 3, Robert, a sixth grader at Roosevelt Middle School, investigated the number of hours of sleep sixth-grade students get on school nights. Today, he is to make a short report to the class on his investigation. Here is his report.

"I took a survey of twenty-nine sixth graders, asking them, 'How many hours of sleep per night do you usually get when you have school the next day?' The first thing I had to do was to organize the data. I did this by drawing a dot plot. Looking at the dot plot, I would say that a typical amount of sleep is 8 or 9 hours."

Dot Plot of Number of Hours of Sleep

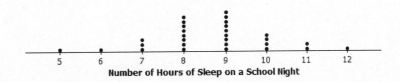

Michelle is Robert's classmate. She liked his report but has a really different thought about determining the center of the number of hours of sleep. Her idea is to even out the data in order to determine a typical or center value.

Exercises 1–6

Suppose that Michelle asks ten of her classmates for the number of hours they usually sleep when there is school the next day.

Suppose they responded (in hours): 8 10 8 8 11 11 9 8 10 7.

1. How do you think Robert would organize this new data? What do you think Robert would say is the center of these ten data points? Why?

2. Do you think his value is a good measure to use for the center of Michelle's data set? Why or why not?

The measure of center that Michelle is proposing is called the *mean*. She finds the total number of hours of sleep for the ten students. That is 90 hours. She has 90 Unifix cubes (Snap cubes). She gives each of the ten students the number of cubes that equals the number of hours of sleep each had reported. She then asks each of the ten students to connect their cubes in a stack and put their stacks on a table to compare them. She then has them share their cubes with each other until they all have the same number of cubes in their stacks when they are done sharing.

3. Make ten stacks of cubes representing the number of hours of sleep for each of the ten students. Using Michelle's method, how many cubes are in each of the ten stacks when they are done sharing?

4. Noting that each cube represents one hour of sleep, interpret your answer to Exercise 3 in terms of number of hours of sleep. What does this number of cubes in each stack represent? What is this value called?

5. Suppose that the student who told Michelle he slept 7 hours changes his data value to 8 hours. What does Michelle's procedure now produce for her center of the new set of data? What did you have to do with that extra cube to make Michelle's procedure work?

6. Interpret Michelle's fair share procedure by developing a mathematical formula that results in finding the fair share value without actually using cubes. Be sure that you can explain clearly how the fair share procedure and the mathematical formula relate to each other.

EUREKA
MATH®

Example 2

Suppose that Robert asked five sixth graders how many pets each had. Their responses were 2, 6, 2, 4, 1. Robert showed the data with cubes as follows:

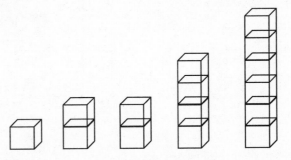

Note that one student has one pet, two students have two pets each, one student has four pets, and one student has six pets. Robert also represented the data set in the following dot plot.

Dot Plot of Number of Pets

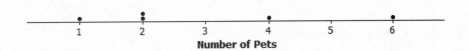

Number of Pets

Robert wants to illustrate Michelle's fair share method by using dot plots. He drew the following dot plot and said that it represents the result of the student with six pets sharing one of her pets with the student who has one pet.

Dot Plot of Number of Pets

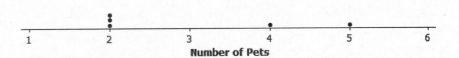

Number of Pets

Robert also represented the dot plot above with cubes. His representation is shown below.

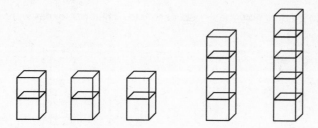

Exercises 7–10

Now, continue distributing the pets based on the following steps.

7. Robert does a fair share step by having the student with five pets share one of her pets with one of the students with two pets.

    a. Draw the cubes representation that shows the result of this fair share step.

    b. Draw the dot plot that shows the result of this fair share step.

8. Robert does another fair share step by having one of the students who has four pets share one pet with one of the students who has two pets.

    a. Draw the cubes representation that shows the result of this fair share step.

b. Draw the dot plot that shows the result of this fair share step.

9. Robert does a final fair share step by having the student who has four pets share one pet with the student who has two pets.

   a. Draw the cubes representation that shows the result of this final fair share step.

   b. Draw the dot plot representation that shows the result of this final fair share step.

10. Explain in your own words why the final representations using cubes and a dot plot show that the mean number of pets owned by the five students is 3 pets.

Name _____   Date _____

1. If a class of 27 students had a mean of 72 on a test, interpret the mean of 72 in the sense of a fair share measure of the center of the test scores.

2. Suppose that your school's soccer team has scored a mean of 2 goals in each of 5 games.

    a. Draw a representation using cubes that displays that your school's soccer team has scored a mean of 2 goals in each of 5 games. Let 1 cube stand for 1 goal.

    b. Draw a dot plot that displays that your school's soccer team has scored a mean of 2 goals in each of 5 games.

1. A game is played where ten tennis balls are tossed into a bucket from a specific distance. The numbers of successful tosses for five students are 7, 2, 5, 1, 5.

   a. Draw a representation of the data using cubes where one cube represents one successful toss of a tennis ball into the bucket.

   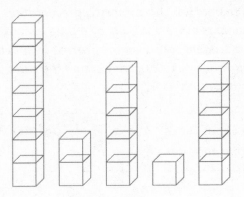

   b. Represent the original data set using a dot plot.

   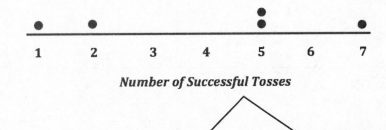

   Number of Successful Tosses

   In the data set, there is one student each who successfully tossed the balls 1, 2, and 7 times, so I can place one dot above each of those numbers in the dot plot to represent each student. There are two students who successfully tossed the ball 5 times each, so I can place two dots above the 5 to represent these two students.

EUREKA
MATH®

2. Find the mean number of successful tosses for this data set using the fair share method. For each step, show the cubes representation and the corresponding dot plot. Explain each step in words in the context of the problem. You may move more than one successful toss in a step, but be sure your explanation is clear. You must show two or more steps.

There are several ways of getting to the final fair share cubes representation where each of the five stacks contains four cubes. Ideally, students move one cube at a time because the leveling is seen more easily that way for many students. If a student shortcuts the process by moving several cubes at once, that is okay, as long as the graphic representations are correct and the explanation is clear. The table below provides one possible representation.

Step Described in Words	Fair Share Cubes Representation	Dot Plot	
Share two of the cubes in the 7-cube stack with the 2-cube stack. The result would be 5, 4, 5, 1, 5. The 7-cube stack went from 7 successful tosses to 5 successful tosses, and the 2-cube stack went from 2 successful tosses to 4 successful tosses.			I can refer back to the original data set to help me understand the moves in the first step.
Then, two of the students who have 5 successful tosses share 1 toss with the student who had 1 successful toss. Those two students with 5 successful tosses went down 1 toss each to 4 successful tosses, and the student with 1 successful toss went up 2 tosses to 3 successful tosses. The result would be 4, 4, 4, 3, 5.			I can act this out with actual cubes, coins, paper clips, or anything I find around the house that can help me visualize these steps.
Finally, the last student with 5 successful tosses shares one of them with the student who has 3 successful tosses. The final step of the fair share method shows an even number of tosses for each of the five students. So, the mean number of successful tosses for these five students is 4 tosses.			

Lesson 6:   Describing the Center of a Distribution Using the Mean

EUREKA MATH®

3. The numbers of granola bars six students brought to school today are 1, 2, 3, 1, 4, 4. Casey produces the following cubes representation as she does the fair share process. Help her decide how to finish the process now that she has stacks of 2, 2, 2, 2, 3, 4.

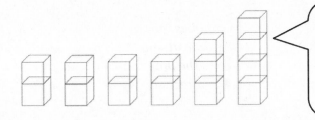

To get to this step, I can share one cube from the 3-cube stack with a 1-cube stack. I can also share one cube from one of the 4-cube stacks with the other 1-cube stack.

There are three extra cubes within the stacks of three and four cubes. Since there are six stacks, each extra cube will need to be split in half so that there are six halves. Each of the six stacks will then have a total of two and one half cubes. In the context of the problem, each student will have a fair share mean of two and one half granola bars.

4. Suppose that the mean number of blueberries in 15 blueberry pancakes is 8 blueberries.

   a. Interpret the mean number of blueberries in terms of fair share.

   Answers will vary. If each of the 15 blueberry pancakes were to have the same number of blueberries, each would have 8 blueberries.

   b. Describe the dot plot representation of the fair share mean of 8 blueberries in 15 pancakes.

   Answers will vary. There should be 15 dots on the dot plot, all of them stacked up at 8.

1. A game was played where ten tennis balls are tossed into a basket from a certain distance. The numbers of successful tosses for six students were 4, 1, 3, 2, 1, 7.

    a. Draw a representation of the data using cubes where one cube represents one successful toss of a tennis ball into the basket.

    b. Represent the original data set using a dot plot.

2. Find the mean number of successful tosses for this data set using the fair share method. For each step, show the cubes representation and the corresponding dot plot. Explain each step in words in the context of the problem. You may move more than one successful toss in a step, but be sure that your explanation is clear. You must show two or more steps.

Step Described in Words	Fair Share Cubes Representation	Dot Plot

3. The numbers of pockets in the clothes worn by four students to school today are 4, 1, 3, and 6. Paige produces the following cubes representation as she does the fair share process. Help her decide how to finish the process now that she has stacks of 3, 3, 3, and 5 cubes.

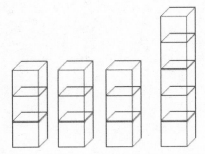

4. Suppose that the mean number of chocolate chips in 30 cookies is 14 chocolate chips.

   a. Interpret the mean number of chocolate chips in terms of fair share.

   b. Describe the dot plot representation of the fair share mean of 14 chocolate chips in 30 cookies.

5. Suppose that the following are lengths (in millimeters) of radish seedlings grown in identical conditions for three days: 12 11 12 14 13 9 13 11 13 10 10 14 16 13 11.

   a. Find the mean length for these 15 radish seedlings.

   b. Interpret the value from part (a) in terms of the fair share mean length.

EUREKA MATH®

In Lesson 3, Robert gave us an informal interpretation of the center of a data set. In Lesson 6, Michelle developed a more formal interpretation of center as a fair share mean, a value that every person in the data set would have if they all had the same value. In this lesson, Sabina will show us how to interpret the mean as a balance point.

Sabina wants to know how long it takes students to get to school. She asks two students how long it takes them to get to school. It takes one student 1 minute and the other student 11 minutes. Sabina represents these data values on a ruler, putting a penny at 1 inch and another at 11 inches. Sabina thinks that there might be a connection between the mean of two data points and where they balance on a ruler. She thinks the mean may be the balancing point. Sabina shows her data using a dot plot.

Dot Plot of Number of Minutes

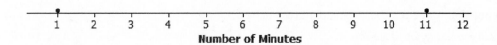

Number of Minutes

Sabina decides to move the penny at 1 inch to 4 inches and the other penny from 11 inches to 8 inches on the ruler, noting that the movement for the two pennies is the same distance but in opposite directions. Sabina thinks that if two data points move the same distance but in opposite directions, the balancing point on the ruler does not change. Do you agree with Sabina?

Sabina continues by moving the penny at 4 inches to 6 inches. To keep the ruler balanced at 6 inches, how far should Sabina move the penny from 8 inches, and in what direction?

Exercises 1–2

Now it is your turn to try balancing two pennies on a ruler.

1. Tape one penny at 2.5 inches on your ruler.

    a. Where should a second penny be taped so that the ruler will balance at 6 inches?

    b. How far is the penny at 2.5 inches from 6 inches? How far is the other penny from 6 inches?

    c. Is 6 inches the mean of the two locations of the pennies? Explain how you know this.

2. Move the penny that is at 2.5 inches to the right two inches.

    a. Where will the penny be placed?

    b. What do you have to do with the other data point (the other penny) to keep the balance point at 6 inches?

    c. What is the mean of the two new data points? Is it the same value as the balance point of the ruler?

EUREKA
MATH

Example 2: Balancing More Than Two Points

Sabina wants to know what happens if there are more than two data points. Suppose there are three students. One student lives 2 minutes from school, and another student lives 9 minutes from school. If the mean time for all three students is 6 minutes, she wonders how long it takes the third student to get to school. Using what you know about distances from the mean, where should the third penny be placed in order for the mean to be 6 inches? Label the diagram, and explain your reasoning.

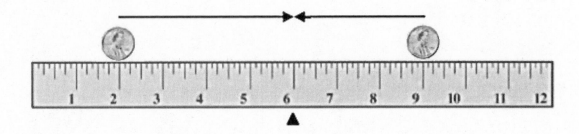

Exercises 3–6

Imagine you are balancing pennies on a ruler.

3.   Suppose you place one penny each at 3 inches, 7 inches, and 8 inches on your ruler.

   a.   Sketch a picture of the ruler. At what value do you think the ruler will balance? Mark the balance point with the symbol Δ.

   b.   What is the mean of 3 inches, 7 inches, and 8 inches? Does your ruler balance at the mean?

c.   Show the information from part (a) on a dot plot. Mark the balance point with the symbol Δ.

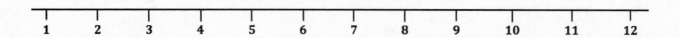

d.   What are the distances on each side of the balance point? How does this prove the mean is 6?

4.   Now, suppose you place a penny each at 7 inches and 9 inches on your ruler.

a.   Draw a dot plot representing these two pennies.

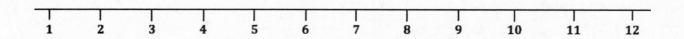

b.   Estimate where to place a third penny on your ruler so that the ruler balances at 6, and mark the point on the dot plot above. Mark the balancing point with the symbol Δ.

c.   Explain why your answer in part (b) is true by calculating the distances of the points from 6. Are the totals of the distances on either side of the mean equal?

EUREKA
MATH®

5.   Is the concept of the mean as the balance point true if you put multiple pennies on a single location on the ruler?

6.   Suppose you place two pennies at 7 inches and one penny at 9 inches on your ruler.

   a.   Draw a dot plot representing these three pennies.

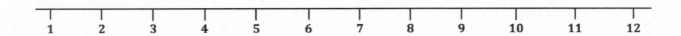

   b.   Estimate where to place a fourth penny on your ruler so that the ruler balances at 6, and mark the point on the dot plot above. Mark the balance point with the symbol Δ.

   c.   Explain why your answer in part (b) is true by calculating the distances of the points from 6. Are the totals of the distances on either side of the mean equal?

Example 3: Finding the Mean

What if the data on a dot plot were 1, 3, and 8? Will the data balance at 6? If not, what is the balance point, and why?

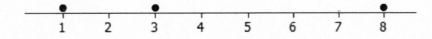

Exercise 7

Use what you have learned about the mean to answer the following questions.

7. Recall from Lesson 6 that Michelle asked ten of her classmates for the number of hours they usually sleep when there is school the next day. Their responses (in hours) were 8, 10, 8, 8, 11, 11, 9, 8, 10, 7.

    a. It's hard to balance ten pennies. Instead of actually using pennies and a ruler, draw a dot plot that represents the data set.

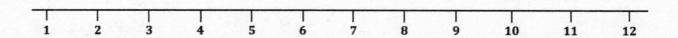

    b. Use your dot plot to find the balance point.

EUREKA MATH

Name _____   Date _____

The dot plot below shows the number of goals scored by a school's soccer team in 7 games so far this season.

Number of Goals Scored

Use the balancing process to explain why the mean number of goals scored is 3.

1. The number of pencils brought to school today by four students is 5, 5, 4, and 2.

   a. Perform the fair share process to find the mean number of pencils for these four students. Sketch the cubes representations for each step of the process.

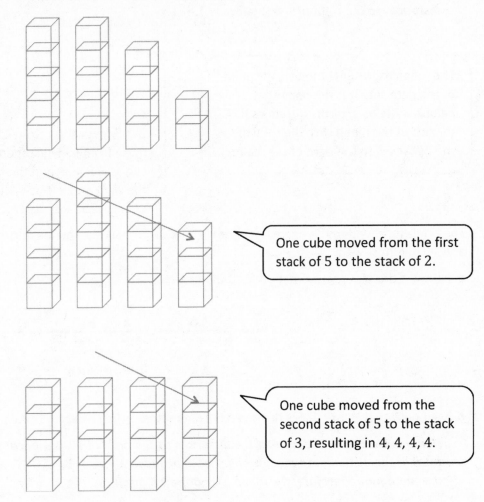

One cube moved from the first stack of 5 to the stack of 2.

One cube moved from the second stack of 5 to the stack of 3, resulting in 4, 4, 4, 4.

Each of the students with 5 pencils gives one pencil to the student who started with 2 pencils, yielding four students with four pencils each. Moving the cubes should result in 4 cubes in each of the four stacks. The mean is 4 pencils.

EUREKA
MATH®

b. Find the total of the distances on each side of the mean to show the mean found in part (a) is correct.

The mean is correct because the total of the distances to the left of 4 *is* 2*, and the total of the distances to the right of* 4 *is* 2 *because* $1 + 1 = 2$.

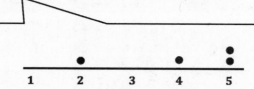

The mean represents the balance point of the data set. It is the point that balances the total of the distances to the left of the mean with the total of the distances to the right of the mean.

2 is 2 units away from 4 (the mean). 5 is 1 unit away from 4. Since there are two 5's in the data set, $1 + 1 = 2$, and the total of the distances to the right and left of 4 are both 2.

2. The times (rounded to the nearest minute) it took each of six classmates to run one mile are 9, 9, 11, 13, 14, and 16 minutes.

a. Draw a dot plot representation for the mile times.

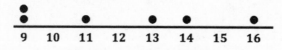

1-Mile Run Times (minutes)

b. Suppose that Henry thinks the mean is 13 minutes. Is he correct? Explain your answer.

Henry is incorrect. The total of the distances to the right of 13 *is* 4 *because* $1 + 3 = 4$*, and the total of the distances to the left of* 13 *is* 10 *because* $2 + 4 + 4 = 10$*. The totals of the distances are not equal; therefore, the mean cannot be* 13 *minutes.*

11 is 2 units to the left of 13, and 9 is 4 units to the left of 13. There are two 9's in the data set, so $2 + 4 + 4 = 10$.

14 is 1 unit to the right of 13, and 16 is 3 units to the right of 13, so $1 + 3 = 4$.

c. What is the mean?

For the total of the distances to be equal on either side of the mean, the mean must be 12 *because the total of the distances to the left of* 12 *is* 7 *because* $1 + 3 + 3 = 7$*, and the total of the distances to the right of* 12 *is* 7 *because* $1 + 2 + 4 = 7$.

EUREKA MATH

3. The number of computers (laptop and desktop) owned by the members of each of eleven families is 1, 2, 2, 3, 5, 5, 6, 6, 7, 7, 11.

   a. Use the mathematical formula for the mean (determine the sum of the data points, and divide by the number of data points) to find the mean number of computers owned for these eleven families.

   $$\frac{55}{11} = 5$$

   The mean is 5 computers.

   b. Draw a dot plot of the data, and verify your answer in part (a) by using the balancing process.

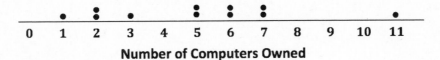

Number of Computers Owned

   The total of the distances to the left of 5 is 12 because $2 + 3 + 3 + 4 = 12$. The total of the distances to the right of 5 is 12 because $1 + 1 + 2 + 2 + 6 = 12$. Since both totals are equal, 5 is the correct mean.

1. The number of pockets in the clothes worn by four students to school today is 4, 1, 3, 4.

    a. Perform the fair share process to find the mean number of pockets for these four students. Sketch the cubes representations for each step of the process.

    b. Find the total of the distances on each side of the mean to show the mean found in part (a) is correct.

2. The times (rounded to the nearest minute) it took each of six classmates to run a mile are 7, 9, 10, 11, 11, and 12 minutes.

    a. Draw a dot plot representation for the mile times.

    b. Suppose that Sabina thinks the mean is 11 minutes. Is she correct? Explain your answer.

    c. What is the mean?

3. The prices per gallon of gasoline (in cents) at five stations across town on one day are shown in the following dot plot. The price for a sixth station is missing, but the mean price for all six stations was reported to be 380 cents per gallon. Use the balancing process to determine the price of a gallon of gasoline at the sixth station.

Dot Plot of Price (cents per gallon)

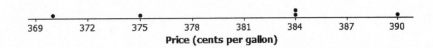

Price (cents per gallon)

4. The number of phones (landline and cell) owned by the members of each of nine families is 3, 5, 6, 6, 6, 6, 7, 7, 8.

    a. Use the mathematical formula for the mean (determine the sum of the data points, and divide by the number of data points) to find the mean number of phones owned for these nine families.

    b. Draw a dot plot of the data, and verify your answer in part (a) by using the balancing process.

Example 1: Comparing Two Data Distributions

Robert's family is planning to move to either New York City or San Francisco. Robert has a cousin in San Francisco and asked her how she likes living in a climate as warm as San Francisco. She replied that it doesn't get very warm in San Francisco. He was surprised by her answer. Because temperature was one of the criteria he was going to use to form his opinion about where to move, he decided to investigate the temperature distributions for New York City and San Francisco. The table below gives average temperatures (in degrees Fahrenheit) for each month for the two cities.

City	Jan.	Feb.	Mar.	Apr.	May	June	July	Aug.	Sep.	Oct.	Nov.	Dec.
New York City	39	42	50	61	71	81	85	84	76	65	55	47
San Francisco	57	60	62	63	64	67	67	68	70	69	63	58

Data Source as of 2013: http://www.usclimatedata.com/climate/san-francisco/california/united-states/usca0987

Data Source as of 2013: http://www.usclimatedata.com/climate/new-york/united-states/3202

Exercises 1–2

Use the data in the table provided in Example 1 to answer the following:

1. Calculate the mean of the monthly average temperatures for each city.

2. Recall that Robert is trying to decide where he wants to move. What is your advice to him based on comparing the means of the monthly temperatures of the two cities?

Example 2: Understanding Variability

Maybe Robert should look at how spread out the New York City monthly temperature data are from the mean of the New York City monthly temperatures and how spread out the San Francisco monthly temperature data are from the mean of the San Francisco monthly temperatures. To compare the variability of monthly temperatures between the two cities, it may be helpful to look at dot plots. The dot plots of the monthly temperature distributions for New York City and San Francisco follow.

Dot Plot of Temperature for New York City

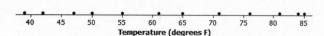

Temperature (degrees F)

Dot Plot of Temperature for San Francisco

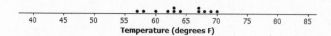

Temperature (degrees F)

Exercises 3–7

Use the dot plots above to answer the following:

3. Mark the location of the mean on each distribution with the balancing Δ symbol. How do the two distributions compare based on their means?

4. Describe the variability of the New York City monthly temperatures from the New York City mean.

5. Describe the variability of the San Francisco monthly temperatures from the San Francisco mean.

Lesson 8:   Variability in a Data Distribution

6. Compare the variability in the two distributions. Is the variability about the same, or is it different? If different, which monthly temperature distribution has more variability? Explain.

7. If Robert prefers to choose the city where the temperatures vary the least from month to month, which city should he choose? Explain.

Example 3: Considering the Mean and Variability in a Data Distribution

The mean is used to describe a typical value for the entire data distribution. Sabina asks Robert which city he thinks has the better climate. How do you think Robert responds?

Sabina is confused and asks him to explain what he means by this statement. How could Robert explain what he means?

Exercises 8–14

Consider the following two distributions of times it takes six students to get to school in the morning and to go home from school in the afternoon.

	Time (minutes)					
Morning	11	12	14	14	16	17
Afternoon	6	10	13	18	18	19

8. To visualize the means and variability, draw a dot plot for each of the two distributions.

Morning

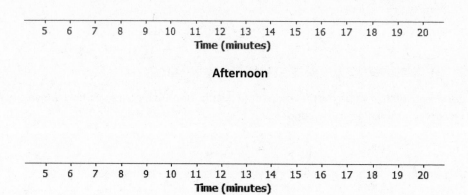

Afternoon

9. What is the mean time to get from home to school in the morning for these six students?

10. What is the mean time to get from school to home in the afternoon for these six students?

11. For which distribution does the mean give a more accurate indicator of a typical time? Explain your answer.

EUREKA MATH®

Distributions can be ordered according to how much the data values vary around their means.

Consider the following data on the number of green jelly beans in seven bags of jelly beans from each of five different candy manufacturers (AllGood, Best, Delight, Sweet, and Yum). The mean in each distribution is 42 green jelly beans.

	Bag 1	Bag 2	Bag 3	Bag 4	Bag 5	Bag 6	Bag 7
AllGood	40	40	41	42	42	43	46
Best	22	31	36	42	48	53	62
Delight	26	36	40	43	47	50	52
Sweet	36	39	42	42	42	44	49
Yum	33	36	42	42	45	48	48

12. Draw a dot plot of the distribution of the number of green jelly beans for each of the five candy makers. Mark the location of the mean on each distribution with the balancing Δ symbol.

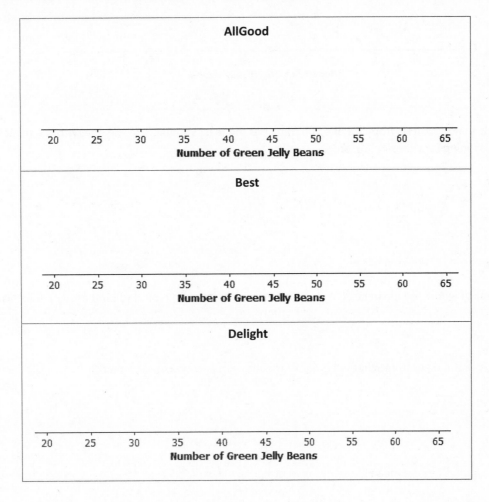

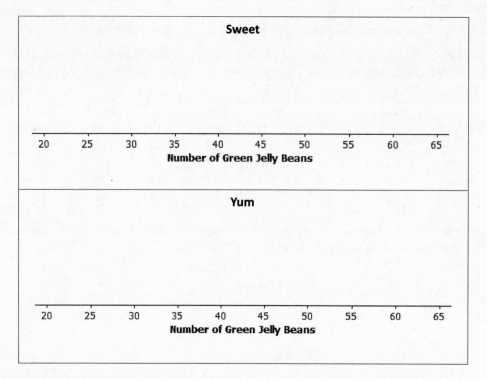

13. Order the candy manufacturers from the one you think has the least variability to the one with the most variability. Explain your reasoning for choosing the order.

14. For which company would the mean be considered a better indicator of a typical value (based on least variability)?

EUREKA
MATH®

Lesson Summary

We can compare distributions based on their means, but variability must also be considered. The mean of a distribution with small variability (not a lot of spread) is considered to be a better indication of a typical value than the mean of a distribution with greater variability (or wide spread).

Name _____   Date _____

1. Consider the following statement: Two sets of data with the same mean will also have the same variability. Do you agree or disagree with this statement? Explain.

2. Suppose the dot plot on the left shows the number of goals a boys' soccer team has scored in 6 games so far this season and the dot plot on the right shows the number of goals a girls' soccer team has scored in 6 games so far this season.

Goals Scored by Boy's Team   Goals Scored by Girl's Team

   a. Compute the mean number of goals for each distribution.

   b. For which distribution, if either, would the mean be considered a better indicator of a typical value? Explain your answer.

1. The number of emails nine employees received in one hour yesterday was 5, 3, 10, 13, 8, 8, 6, 10, and 9. The number of emails received by the same nine employees during the same time period today was 6, 8, 6, 9, 6, 6, 11, 11, and 9.

    a. Draw dot plots for the distributions of the number of emails received yesterday and of the number of emails received today. Be sure to use the same scale on both dot plots.

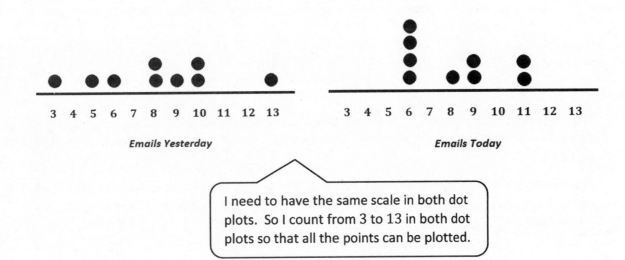

Emails Yesterday

Emails Today

> I need to have the same scale in both dot plots. So I count from 3 to 13 in both dot plots so that all the points can be plotted.

    b. Do the distributions have the same mean? What is the mean of each dot plot?

    Yes, both distributions have a mean of 8 emails.

    > I can use the fair share method that I learned in Lesson 6 to determine the means. Or, I can use the method from Lesson 7, using distances from the center.

2. The following table shows the wait times, in minutes, at five restaurants at 5:00 p.m. across town as recorded on Monday, Wednesday, and Friday of a certain week.

Day	Marla's Diner	Taco, Taco	Tony's Italian Eatery	The Steak House	China Buffet
Monday	5	0	20	30	0
Wednesday	4	4	11	18	18
Friday	11	10	10	15	9

a. The mean wait time per day for the five restaurants is the same for each of the three days. Without doing any calculations and simply looking at Wednesday's wait times, what must the mean wait time be?

Wednesday's times are centered at 11 minutes. The sum of the distances from 11 for values above 11 is equal to the sum of the distances from 11 for values below 11, so the mean is 11 minutes.

b. For which daily distribution is the mean a better indicator of the typical wait time for the five restaurants? Explain.

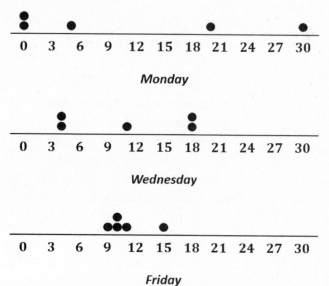

I need to compare the variability in each set to determine if the mean is a good indicator, so I can draw or visualize a dot plot for each to help me. The mean of a set with a smaller spread, where the points are all near the same amount, is considered a better indicator than a set of numbers that have a great amount of variability.

From the dot plots, the mean is the better indicator of the typical wait time for the five restaurants on Friday because there is the least variability in the Friday wait times.

EUREKA MATH

1. The number of pockets in the clothes worn by seven students to school yesterday was 4, 1, 3, 4, 2, 2, 5. Today, those seven students each had three pockets in their clothes.

   a. Draw one dot plot of the number of pockets data for what students wore yesterday and another dot plot for what students wore today. Be sure to use the same scale.

   b. For each distribution, find the mean number of pockets worn by the seven students. Show the means on the dot plots by using the balancing △ symbol.

   c. For which distribution is the mean number of pockets a better indicator of what is typical? Explain.

2. The number of minutes (rounded to the nearest minute) it took to run a certain route was recorded for each of five students. The resulting data were 9, 10, 11, 14, and 16 minutes. The number of minutes (rounded to the nearest minute) it took the five students to run a different route was also recorded, resulting in the following data: 6, 8, 12, 15, and 19 minutes.

   a. Draw dot plots for the distributions of the times for the two routes. Be sure to use the same scale on both dot plots.

   b. Do the distributions have the same mean? What is the mean of each dot plot?

   c. In which distribution is the mean a better indicator of the typical amount of time taken to run the route? Explain.

3. The following table shows the prices per gallon of gasoline (in cents) at five stations across town as recorded on Monday, Wednesday, and Friday of a certain week.

Day	R&C	Al's	PB	Sam's	Ann's
Monday	359	358	362	359	362
Wednesday	357	365	364	354	360
Friday	350	350	360	370	370

   a. The mean price per day for the five stations is the same for each of the three days. Without doing any calculations and simply looking at Friday's prices, what must the mean price be?

   b. For which daily distribution is the mean a better indicator of the typical price per gallon for the five stations? Explain.

Example 1: Variability

In Lesson 8, Robert wanted to decide where he would rather move (New York City or San Francisco). He planned to make his decision by comparing the average monthly temperatures for the two cities. Since the mean of the average monthly temperatures for New York City and the mean for San Francisco turned out to be about the same, he decided instead to compare the cities based on the variability in their monthly average temperatures. He looked at the two distributions and decided that the New York City temperatures were more spread out from their mean than were the San Francisco temperatures from their mean.

Exercises 1–3

The following temperature distributions for seven other cities all have a mean monthly temperature of approximately 63 degrees Fahrenheit. They do not have the same variability.

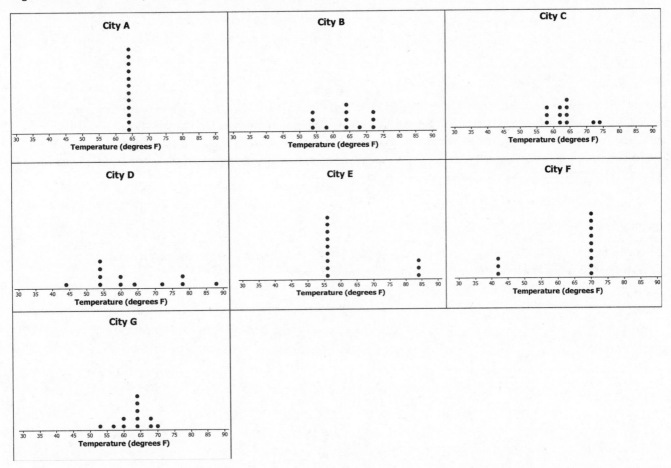

1. Which distribution has the smallest variability? Explain your answer.

2. Which distribution or distributions seem to have the most variability? Explain your answer.

3. Order the seven distributions from least variability to most variability. Explain why you listed the distributions in the order that you chose.

Example 2: Measuring Variability

Based on just looking at the distributions, there are different orderings of variability that seem to make some sense. Sabina is interested in developing a formula that will produce a number that measures the variability in a data distribution. She would then use the formula to measure the variability in each data set and use these values to order the distributions from smallest variability to largest variability. She proposes beginning by looking at how far the values in a data set are from the mean of the data set.

Exercises 4–5

The dot plot for the monthly temperatures in City G is shown below. Use the dot plot and the mean monthly temperature of 63 degrees Fahrenheit to answer the following questions.

City G

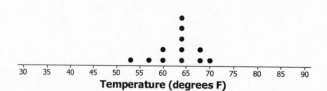

Temperature (degrees F)

Lesson 9:   The Mean Absolute Deviation (MAD)

4. Fill in the following table for City G's temperature deviations.

Temperature (in degrees Fahrenheit)	Distance (in degrees Fahrenheit) from the Mean of 63°F	Deviation from the Mean (distance and direction)
53	10	10 to the left
57		
60		
60		
64		
64		
64		
64		
64		
68		
68		
70		

5. What is the sum of the distances to the left of the mean? What is the sum of the distances to the right of the mean?

EUREKA MATH

Example 3: Finding the Mean Absolute Deviation (MAD)

Sabina notices that when there is not much variability in a data set, the distances from the mean are small and that when there is a lot of variability in a data set, the data values are spread out and at least some of the distances from the mean are large. She wonders how she can use the distances from the mean to help her develop a formula to measure variability.

Exercises 6–7

6. Use the data on monthly temperatures for City G given in Exercise 4 to answer the following questions.

    a. Fill in the following table.

Temperature (in degrees Fahrenheit)	Distance from the Mean (absolute deviation)
53	10
57	
60	
60	
64	
64	
64	
64	
64	
68	
68	
70	

    b. The absolute deviation for a data value is its distance from the mean of the data set. For example, for the first temperature value for City G (53 degrees), the absolute deviation is 10. What is the sum of the absolute deviations?

c. Sabina suggests that the mean of the absolute deviations (the mean of the distances) could be a measure of the variability in a data set. Its value is the average distance of the data values from the mean of the monthly temperatures. It is called the *mean absolute deviation* and is denoted by the letters MAD. Find the MAD for this data set of City G's temperatures. Round to the nearest tenth.

d. Find the MAD values in degrees Fahrenheit for each of the seven city temperature distributions, and use the values to order the distributions from least variability to most variability. Recall that the mean for each data set is 63 degrees Fahrenheit. Looking only at the distributions, does the list that you made in Exercise 2 match the list made by ordering MAD values?

MAD values (in °F):

e. Which of the following is a correct interpretation of the MAD?

    i.   The monthly temperatures in City G are all within 3.7 degrees from the approximate mean of 63 degrees.

    ii. The monthly temperatures in City G are, on average, 3.7 degrees from the approximate mean temperature of 63 degrees.

    iii. All of the monthly temperatures in City G differ from the approximate mean temperature of 63 degrees by 3.7 degrees.

7. The dot plot for City A's temperatures follows.

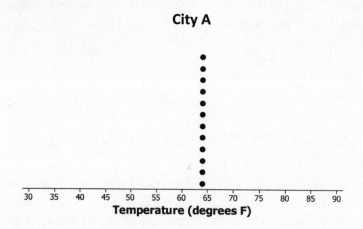

City A

Temperature (degrees F)

a. How much variability is there in City A's temperatures? Why?

b. Does the MAD agree with your answer in part (a)?

EUREKA MATH®

Lesson Summary

In this lesson, a formula was developed that measures the amount of variability in a data distribution.

- The absolute deviation of a data point is the distance that data point is from the mean.

- The mean absolute deviation (MAD) is computed by finding the mean of the absolute deviations (distances from the mean) for the data set.

- The value of MAD is the average distance that the data values are from the mean.

- A small MAD indicates that the data distribution has very little variability.

- A large MAD indicates that the data points are spread out and that at least some are far away from the mean.

Name _____ Date _____

1. The mean absolute deviation (MAD) is a measure of variability for a data set. What does a data distribution look like if its MAD equals zero? Explain.

2. Is it possible to have a negative value for the MAD of a data set?

3. Suppose that seven students have the following numbers of pets: 1, 1, 1, 2, 4, 4, 8.

   a. The mean number of pets for these seven students is 3 pets. Use the following table to find the MAD for this distribution of number of pets.

Student	Number of Pets	Deviation from the Mean (distance and direction)	Absolute Deviation (distance from the mean)
1	1		
2	1		
3	1		
4	2		
5	4		
6	4		
7	8		
Sum			

   b. Explain in words what the MAD means for this data set.

1. Suppose the dot plot on the left shows the number of books a group of friends read last month. The dot plot on the right shows the number of books the same group of friends read two months ago. The mean for both of these groups is 4.

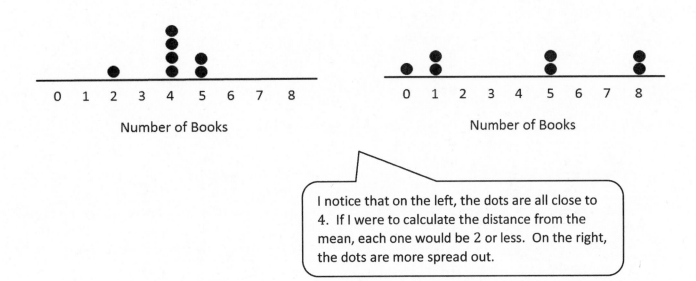

I notice that on the left, the dots are all close to 4. If I were to calculate the distance from the mean, each one would be 2 or less. On the right, the dots are more spread out.

a. Before doing any calculations, which dot plot has the larger MAD? Explain how you know.

The graph showing the books read two months ago has a larger MAD because the data are more spread out and have the larger distances from the mean.

I know that if the points are more spread out on the dot plot, the values are more varied. MAD measures variability. So the farther the dots are on the number line, the larger the MAD.

b. Use the following tables to find the MAD for each distribution. Round your calculations to the nearest hundredth.

Last Month	
Number of Books	**Absolute Deviation**
2	2
4	0
4	0
4	0
4	0
5	1
5	1
Sum	4

Two Months Ago	
Number of Books	**Absolute Deviation**
0	4
1	3
1	3
5	1
5	1
8	4
8	4
Sum	20

I know that absolute deviation is the distance a value is from the mean. The mean in each case is 4, so I just need to know how far each of the points is from 4.

The MAD for last month is about 0.57 books because $\frac{4}{7} \approx 0.57$. The MAD for two months ago is about 2.86 books because $\frac{20}{7} \approx 2.86$.

To determine the MAD, I need to find the sum of all of the absolute deviations and then divide by 7 because there are 7 data values.

2. Consider the following data of the number of light bulbs that do not work in a case of light bulbs sampled from each of five companies. Note that the mean of each distribution is 14 broken light bulbs.

Company	Case 1	Case 2	Case 3	Case 4	Case 5	Case 6
A	24	10	8	12	6	10
B	18	3	9	20	17	3
C	2	36	14	5	5	8
D	16	16	8	12	8	10
E	5	14	14	2	22	13

a. Complete the following table of the absolute deviations for the six cases of light bulbs for each company.

Company	Absolute Deviation					
	Case 1	Case 2	Case 3	Case 4	Case 5	Case 6
A	10	4	6	2	8	4
B	4	11	5	6	3	11
C	12	22	0	9	9	6
D	2	2	6	2	6	4
E	9	0	0	12	8	1

Some of the absolute deviations have been determined for me. I just need to fill in the rest. I need to determine the distance from the mean to the values in the first table.

b. For which company is the mean a better indication of a typical number of broken light bulbs in each case? Explain your answer.

> I know that for the mean to be a better indication of a typical value, the MAD must be small. I will need to calculate the MADs to support my answer.

Company	Absolute Deviation						SUM	MAD
	Case 1	Case 2	Case 3	Case 4	Case 5	Case 6		
A	10	4	6	2	8	4	34	$\frac{34}{6} \approx 5.67$
B	4	11	5	6	3	11	40	$\frac{40}{6} \approx 6.67$
C	12	22	0	9	9	6	58	$\frac{58}{6} \approx 9.67$
D	2	2	6	2	6	4	22	$\frac{22}{6} \approx 3.67$
E	9	0	0	12	8	1	30	$\frac{30}{6} = 5$

> I can extend the absolute deviation table by adding a column for the sum of the absolute deviations and a column for the MAD. This way, I can keep my work organized.

The mean is a good indicator for Company D because the MAD is smaller than for the other companies, showing that there is less variation in the data values.

EUREKA MATH®

1. Suppose the dot plot on the left shows the number of goals a boys' soccer team has scored in six games so far this season, and the dot plot on the right shows the number of goals a girls' soccer team has scored in six games so far this season. The mean for both of these teams is 3.

Dot Plot of Number of Goals Scored for Boys' Team

Dotplot of Number of Goals Scored for Girls' Team

a. Before doing any calculations, which dot plot has the larger MAD? Explain how you know.

b. Use the following tables to find the MAD for each distribution. Round your calculations to the nearest hundredth.

Boys' Team	
Number of Goals	Absolute Deviation
0	
0	
3	
3	
5	
7	
Sum	

Girls' Team	
Number of Goals	Absolute Deviation
2	
2	
3	
3	
3	
5	
Sum	

c. Based on the computed MAD values, for which distribution is the mean a better indication of a typical value? Explain your answer.

2. Recall Robert's problem of deciding whether to move to New York City or to San Francisco. A table of temperatures (in degrees Fahrenheit) and absolute deviations for New York City follows:

Average Temperature in New York City												
Month	Jan.	Feb.	Mar.	Apr.	May	June	July	Aug.	Sep.	Oct.	Nov.	Dec.
Temperature (°F)	39	42	50	61	71	81	85	84	76	65	55	47
Absolute Deviation	24	21	13	2	8	18	22	21	13	2	8	16

a. The absolute deviations for the monthly temperatures are shown in the above table. Use this information to calculate the MAD. Explain what the MAD means in words.

b. Complete the following table, and then use the values to calculate the MAD for the San Francisco data distribution.

Average Temperature in San Francisco												
Month	Jan.	Feb.	Mar.	Apr.	May	June	July	Aug.	Sep.	Oct.	Nov.	Dec.
Temperature (°F)	57	60	62	63	64	67	67	68	70	69	63	58
Absolute Deviation												

c. Comparing the MAD values for New York City and San Francisco, which city would Robert choose to move to if he is interested in having a lot of variability in monthly temperatures? Explain using the MAD.

3. Consider the following data of the number of green jelly beans in seven bags sampled from each of five different candy manufacturers (Awesome, Delight, Finest, Sweeties, YumYum). Note that the mean of each distribution is 42 green jelly beans.

	Bag 1	Bag 2	Bag 3	Bag 4	Bag 5	Bag 6	Bag 7
Awesome	40	40	41	42	42	43	46
Delight	22	31	36	42	48	53	62
Finest	26	36	40	43	47	50	52
Sweeties	36	39	42	42	42	44	49
YumYum	33	36	42	42	45	48	48

a. Complete the following table of the absolute deviations for the seven bags for each candy manufacturer.

Absolute Deviations							
	Bag 1	Bag 2	Bag 3	Bag 4	Bag 5	Bag 6	Bag 7
Awesome	2	2	1	0	0	1	4
Delight	20	11	6				
Finest	16						
Sweeties							
YumYum							

Lesson 9: The Mean Absolute Deviation (MAD)

EUREKA MATH

b. Based on what you learned about MAD, which manufacturer do you think will have the lowest MAD? Calculate the MAD for the manufacturer you selected.

	Bag 1	Bag 2	Bag 3	Bag 4	Bag 5	Bag 6	Bag 7	SUM	MAD
Awesome									
Delight									
Finest									
Sweeties									
YumYum									

© 2019 Great Minds®. eureka-math.org

Example 1: Describing Distributions

In Lesson 9, Sabina developed the mean absolute deviation (MAD) as a number that measures variability in a data distribution. Using the mean and MAD along with a dot plot allows you to describe the center, spread, and shape of a data distribution. For example, suppose that data on the number of pets for ten students are shown in the dot plot below.

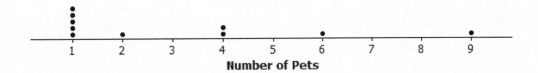

There are several ways to describe the data distribution. The mean number of pets for these students is 3, which is a measure of center. There is variability in the number of pets the students have, and data values differ from the mean by about 2.2 pets on average (the MAD). The shape of the distribution is heavy on the left, and then it thins out to the right.

Exercises 1–4

1. Suppose that the weights of seven middle school students' backpacks are given below.

   a. Fill in the following table.

Student	Alan	Beth	Char	Damon	Elisha	Fred	Georgia
Weight (pounds)	18	18	18	18	18	18	18
Deviation							
Absolute Deviation							

   b. Draw a dot plot for these data, and calculate the mean and MAD.

EUREKA
MATH®

   c.   Describe this distribution of weights of backpacks by discussing the center, spread, and shape.

2.   Suppose that the weight of Elisha's backpack is 17 pounds rather than 18 pounds.

   a.   Draw a dot plot for the new distribution.

   b.   Without doing any calculations, how is the mean affected by the lighter weight? Would the new mean be the same, smaller, or larger?

   c.   Without doing any calculations, how is the MAD affected by the lighter weight? Would the new MAD be the same, smaller, or larger?

3.   Suppose that in addition to Elisha's backpack weight having changed from 18 to 17 pounds, Fred's backpack weight is changed from 18 to 19 pounds.

   a.   Draw a dot plot for the new distribution.

EUREKA
MATH

b. Without doing any calculations, how would the new mean compare to the original mean?

c. Without doing any calculations, would the MAD for the new distribution be the same as, smaller than, or larger than the original MAD?

d. Without doing any calculations, how would the MAD for the new distribution compare to the one in Exercise 2?

4. Suppose that seven second graders' backpack weights were as follows:

Student	Alice	Bob	Carol	Damon	Ed	Felipe	Gale
Weight (pounds)	5	5	5	5	5	5	5

a. How is the distribution of backpack weights for the second graders similar to the original distribution for the middle school students given in Exercise 1?

b. How are the distributions different?

Example 2: Using the MAD

Using data to make decisions often involves comparing distributions. Recall that Robert is trying to decide whether to move to New York City or to San Francisco based on temperature. Comparing the center, spread, and shape for the two temperature distributions could help him decide.

Dot Plot of Temperature for New York City   **Dot Plot of Temperature for San Francisco**

From the dot plots, Robert saw that monthly temperatures in New York City were spread fairly evenly from around 40 degrees to around 85 degrees, but in San Francisco, the monthly temperatures did not vary as much. He was surprised that the mean temperature was about the same for both cities. The MAD of 14 degrees for New York City told him that, on average, a month's temperature was 14 degrees away from the mean of 63 degrees. That is a lot of variability, which is consistent with the dot plot. On the other hand, the MAD for San Francisco told him that San Francisco's monthly temperatures differ, on average, only 3.5 degrees from the mean of 64 degrees. So, the mean doesn't help Robert very much in making a decision, but the MAD and dot plot are helpful.

Which city should he choose if he loves warm weather and really dislikes cold weather?

Exercises 5–7

5. Robert wants to compare temperatures in degrees Fahrenheit for Cities B and C.

	Jan.	Feb.	Mar.	Apr.	May	June	July	Aug.	Sept.	Oct.	Nov.	Dec.
City B	54	54	58	63	63	68	72	72	72	63	63	54
City C	54	44	54	61	63	72	78	85	78	59	54	54

a. Draw a dot plot of the monthly temperatures for each of the cities.

EUREKA
MATH

b. Verify that the mean monthly temperature for each distribution is 63 degrees.

c. Find the MAD for each of the cities. Interpret the two MADs in words, and compare their values. Round your answers to the nearest tenth of a degree.

6. How would you describe the differences in the shapes of the monthly temperature distributions of the two cities?

7. Suppose that Robert had to decide between Cities D, E, and F.

	Jan.	Feb.	Mar.	Apr.	May	June	July	Aug.	Sept.	Oct.	Nov.	Dec.	Mean	MAD
City D	54	44	54	59	63	72	78	87	78	59	54	54	63	10.5
City E	56	56	56	56	56	84	84	84	56	56	56	56	63	10.5
City F	42	42	70	70	70	70	70	70	70	70	70	42	63	10.5

a. Draw a dot plot for each distribution.

b.   Interpret the MAD for the distributions. What does this mean about variability?

c.   How will Robert decide to which city he should move? List possible reasons Robert might have for choosing each city.

Lesson Summary

A data distribution can be described in terms of its center, spread, and shape.

- The center can be measured by the mean.
- The spread can be measured by the mean absolute deviation (MAD).
- A dot plot shows the shape of the distribution.

Name _____   Date _____

1. A dot plot of times that five students studied for a test is displayed below.

Studying for a Test

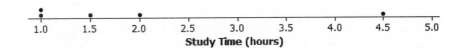

Study Time (hours)

a. Calculate the mean number of hours that these five students studied. Then, use the mean to calculate the absolute deviations, and complete the table.

Student	Aria	Ben	Chloe	Dellan	Emma
Number of Study Hours	1	1	1.5	2	4.5
Absolute Deviation					

b. Find and interpret the MAD for this data set.

2. The same five students are preparing to take a second test. Suppose that the numbers of study hours were the same except that Ben studied 2.5 hours for the second test (1.5 hours more), and Emma studied only 3 hours for the second test (1.5 hours less).

a. Without doing any calculations, is the mean for the second test the same as, greater than, or less than the mean for the first test? Explain your reasoning.

b. Without doing any calculations, is the MAD for the second test the same as, greater than, or less than the MAD for the first test? Explain your reasoning.

Seven people were timed while completing an obstacle course. Their finishing times are shown in the table.

Person	Adam	Bertram	Corrine	Diego	Enrique	Frieda	Gretchen
Time (in minutes)	6	7.5	5	6	7	6	4.5

   a. Draw a dot plot for the finishing times for these seven participants.

Obstacle Course Finishing Times

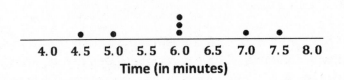

I remember that I need to label my dot plot. I can use my dot plot to see if my data set is symmetrical or if the data clusters around one value.

   b. Find the mean finishing time for the seven participants.

$$\frac{(6 + 7.5 + 5 + 6 + 7 + 6 + 4.5)}{7} = \frac{42}{7} = 6$$

The mean of the data is 6 minutes.

c. Corrine said that the MAD for this data set is 0 minutes because the dot plot is balanced around 6. Without doing any calculations, do you agree with Corrine? Why or why not? If not, calculate the MAD, and state what it means.

No, Corrine is wrong. There is variability in the data. Not all points are on 6.

> In Lesson 9, I learned that I must determine the sum of the absolute deviations and then divide by the number of data values to calculate the MAD.

> If all the participants finished at the same time, the MAD would be 0. If this were the case, my dot plot would show dots only at 6 and no other finishing times.

The sum of the absolute deviations is 5. So, $\frac{5}{8} = 0.625$; therefore the MAD is 0.625 minutes. This means that, on average, the number of minutes these participants took to complete the obstacle course on a typical day differs by 0.625 minutes from the group mean of 6 minutes.

d. Suppose that in the original data set, Adam needs an additional two minutes to complete the course, and Frieda needs two less minutes to complete the obstacle course.

i.   Without doing any calculations, does the mean for the new data set stay the same, increase, or decrease as compared to the original mean? Explain your reasoning.

The mean would remain at 6 minutes. One data value moved the same number of units to the right as another data value moved to the left, so, the balance point of the distribution does not change.

ii. Without doing any calculations, does the MAD for the new data set stay the same, increase, or decrease as compared to the original MAD? Explain your reasoning.

Since both scores moved away from the mean, the resulting distribution would be more spread out than the original distribution. Therefore, the MAD would increase.

> If I think about the sum of the absolute deviations in part (c), I added in zeros for Adam, Diego, and Frieda because their times were the same as the mean. Now that Adam's and Frieda's time will change, I will have to add more to the sum; therefore, the MAD will increase too.

EUREKA MATH®

1. Draw a dot plot of the times that five students studied for a test if the mean time they studied was 2 hours and the MAD was 0 hours.

2. Suppose the times that five students studied for a test are as follows:

Student	Aria	Ben	Chloe	Dellan	Emma
Time (hours)	1.5	2	2	2.5	2

   Michelle said that the MAD for this data set is 0 hours because the dot plot is balanced around 2. Without doing any calculations, do you agree with Michelle? Why or why not?

3. Suppose that the number of text messages eight students receive on a typical day is as follows:

Student	1	2	3	4	5	6	7	8
Number of Text Messages	42	56	35	70	56	50	65	50

   a. Draw a dot plot for the number of text messages received on a typical day for these eight students.

   b. Find the mean number of text messages these eight students receive on a typical day.

   c. Find the MAD for the number of text messages, and explain its meaning using the words of this problem.

   d. Describe the shape of this data distribution.

   e. Suppose that in the original data set, Student 3 receives an additional five text messages per day, and Student 4 receives five fewer text messages per day.

   i. Without doing any calculations, does the mean for the new data set stay the same, increase, or decrease as compared to the original mean? Explain your reasoning.

   ii. Without doing any calculations, does the MAD for the new data set stay the same, increase, or decrease as compared to the original MAD? Explain your reasoning.

Example 1: Comparing Distributions with the Same Mean

In Lesson 10, a data distribution was characterized mainly by its center (mean) and variability (MAD). How these measures help us make a decision often depends on the context of the situation. For example, suppose that two classes of students took the same test, and their grades (based on 100 points) are shown in the following dot plots. The mean score for each distribution is 79 points. Would you rather be in Class A or Class B if you had a score of 79?

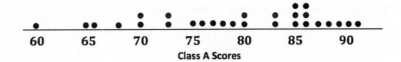

Class A Scores

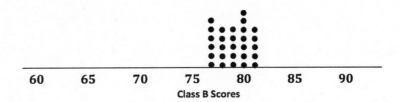

Class B Scores

Exercises 1–6

1. Looking at the dot plots, which class has the greater MAD? Explain without actually calculating the MAD.

2. If Liz had one of the highest scores in her class, in which class would she rather be? Explain your reasoning.

3. If Logan scored below average, in which class would he rather be? Explain your reasoning.

Your little brother asks you to replace the battery in his favorite remote control car. The car is constructed so that it is difficult to replace its battery. Your research of the lifetimes (in hours) of two different battery brands (A and B) shows the following lifetimes for 20 batteries from each brand:

A	12	14	14	15	16	17	17	18	19	20	21	21	23	23	24	24	24	25	26	27
B	18	18	19	19	19	19	19	19	20	20	20	20	20	21	21	21	21	22	22	22

4. To help you decide which battery to purchase, start by drawing a dot plot of the lifetimes for each brand.

5. Find the mean battery lifetime for each brand, and compare them.

6. Looking at the variability in the dot plot for each data set, give one reason you might choose Brand A. What is one reason you might choose Brand B? Explain your reasoning.

Example 2: Comparing Distributions with Different Means

You have been comparing distributions that have the same mean but different variability. As you have seen, deciding whether large variability or small variability is best depends on the context and on what is being asked. If two data distributions have different means, do you think that variability will still play a part in making decisions?

EUREKA
MATH

Exercises 7–9

Suppose that you wanted to answer the following question: Are field crickets better predictors of air temperature than katydids? Both species of insect make chirping sounds by rubbing their front wings together.

The following data are the number of chirps (per minute) for 10 insects of each type. All the data were taken on the same evening at the same time.

Insect	1	2	3	4	5	6	7	8	9	10
Crickets	35	32	35	37	34	34	38	35	36	34
Katydids	66	62	61	64	63	62	68	64	66	64

7. Draw dot plots for these two data distributions using the same scale, going from 30 to 70. Visually, what conclusions can you draw from the dot plots?

8. Calculate the mean and MAD for each distribution.

9. The outside temperature T, in degrees Fahrenheit, can be predicted by using two different formulas. The formulas include the mean number of chirps per minute made by crickets or katydids.

a. For crickets, T is predicted by adding 40 to the mean number of chirps per minute. What value of T is being predicted by the crickets?

b. For katydids, T is predicted by adding 161 to the mean number of chirps per minute and then dividing the sum by 3. What value of T is being predicted by the katydids?

c. The temperature was 75 degrees Fahrenheit when these data were recorded, so using the mean from each data set gave an accurate prediction of temperature. If you were going to use the number of chirps from a single cricket or a single katydid to predict the temperature, would you use a cricket or a katydid? Explain how variability in the distributions of number of chirps played a role in your decision.

Lesson 11:   Describing Distributions Using the Mean and MAD

© 2019 Great Minds®. eureka-math.org

Name _____   Date _____

You need to decide which of two brands of chocolate chip cookies to buy. You really love chocolate chip cookies. The numbers of chocolate chips in each of five cookies from each brand are as follows:

Cookie	1	2	3	4	5
ChocFull	17	19	18	18	18
AllChoc	22	15	14	21	18

a. Draw a dot plot for each set of data that shows the distribution of the number of chips for that brand. Use the same scale for both of your dot plots (one that covers the span of both distributions).

b. Find the mean number of chocolate chips for each of the two brands. Compare the means.

c. Looking at your dot plots and considering variability, which brand do you prefer? Explain your reasoning.

1. Two bowling teams wrote down their scores for the last game. Summary measures for the two teams are as follows:

	Mean	MAD
Team A	184	15
Team B	184	4

> I know that a larger MAD shows more variability. That means that the scores will possibly range from much lower to much higher than the mean for Team A. The small MAD for Team B means that the scores are clustered closer together.

a. Suppose that Rita, the newest bowler, scored the lowest score on the team. Would she have scored lower on Team A or Team B? Explain your reasoning.

Rita's score would have been lower if she had been on Team A because the means are the same, and the variability, as measured by the MAD, is higher on that team than it is for Team B. This tells me that the lowest score for Team A will be lower than the lowest score for Team B.

b. Suppose that your score was below the mean score. On which team would you prefer to have been? Explain your reasoning.

> Because the MAD is larger for Team A, a score below the mean score could be a lot lower than the mean. Because the MAD is smaller for Team B, a score below the mean score would be much closer to the mean.

I would prefer to have been on Team B because then my score would most likely be closer to the mean of 184 than if I were on Team A. A score below the mean on Team A could be far lower than on Team B because the MAD of Team A is much larger.

2. A beekeeper keeps 8 hives in each of two of his apiaries, or bee yards. The numbers of pounds of honey produced by each hive is shown:

	Hive	1	2	3	4	5	6	7	8
Pounds of Honey	Yard A	40	42	57	45	52	73	45	78
	Yard B	43	50	45	46	44	46	44	50

a. Draw dot plots to help you decide which bee yard is more productive. Use the same scale for both of your dot plots (one that covers the span of both distributions).

Hive Productivity in Yard A

40   45   50   55   60   65   70   75   80
Pounds of Honey

> I need to determine the smallest and largest numbers in the table to determine the range of numbers to use in my dot plots.

Hive Productivity in Yard B

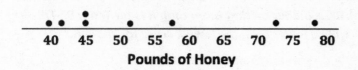

40   45   50   55   60   65   70   75   80
Pounds of Honey

b. Calculate the mean number of pounds of honey produced for each bee yard. Which one produces more pounds of honey on average?

*The mean number of pounds for Yard A is **54** because* $\dfrac{40 + 42 + 57 + 45 + 52 + 73 + 45 + 78}{8} = 54$*, and the mean number of pounds for Yard B is **46** because* $\dfrac{43 + 50 + 45 + 46 + 44 + 46 + 44 + 50}{8} = 46$*. Yard A produces more pounds of honey on average.*

> I can use the values in the table at the beginning of the problem to determine the mean.

Lesson 11:   Describing Distributions Using the Mean and MAD

EUREKA MATH

c. If you want to be able to accurately predict the number of pounds of honey a bee yard will produce, which yard should you choose—the one with the smaller MAD or the one with the larger MAD? Explain your reasoning.

Yard B produces fewer pounds of honey on average but is far more consistent. Looking at the dot plots, its variability is far less than that of Yard A. Based on these data sets, choosing Yard B should yield numbers in the mid 40's consistently, but the numbers from Yard A could vary wildly from the low 40's to huge yields around 80.

d. Calculate the MAD of each bee yard.

	Hive	1	2	3	4	5	6	7	8
Absolute Deviations (in pounds)	Yard A	14	12	3	9	2	19	9	24
	Yard B	3	4	1	0	2	0	2	4

> I can make a table of the absolute deviations. This will help me keep my data organized. I also need to remember that the means are different. In Yard A, the mean is 54 pounds, and in Yard B, the mean is 46 pounds.

The sum of the distances from the mean for Yard A is 92 because $14 + 12 + 3 + 9 + 2 + 19 + 9 + 24 = 92$. Therefore, the MAD for Yard A is 11.5 pounds because $\frac{92}{8} = 11.5$.

The sum of the distances from the mean for Yard B is 16 because $3 + 4 + 1 + 0 + 2 + 0 + 2 + 4 = 16$. Therefore, the MAD for Yard B is 2 pounds because $\frac{16}{8} = 2$.

1. Two classes took the same mathematics test. Summary measures for the two classes are as follows:

	Mean	MAD
Class A	78	2
Class B	78	10

   a. Suppose that you received the highest score in your class. Would your score have been higher if you were in Class A or Class B? Explain your reasoning.

   b. Suppose that your score was below the mean score. In which class would you prefer to have been? Explain your reasoning.

2. Eight of each of two varieties of tomato plants, LoveEm and Wonderful, are grown under the same conditions. The numbers of tomatoes produced from each plant of each variety are shown:

Plant	1	2	3	4	5	6	7	8
LoveEm	27	29	27	28	31	27	28	27
Wonderful	31	20	25	50	32	25	22	51

   a. Draw dot plots to help you decide which variety is more productive.

   b. Calculate the mean number of tomatoes produced for each variety. Which one produces more tomatoes on average?

   c. If you want to be able to accurately predict the number of tomatoes a plant is going to produce, which variety should you choose—the one with the smaller MAD or the one with the larger MAD? Explain your reasoning.

   d. Calculate the MAD of each plant variety.

How do we summarize a data distribution? What provides us with a good description of the data? The following exercises help us to understand how a numerical summary provides an answer to these questions.

Example 1: The Median—A Typical Number

Suppose a chain restaurant (Restaurant A) advertises that a typical number of french fries in a large bag is 82. The dot plot shows the number of french fries in a sample of twenty large bags from Restaurant A.

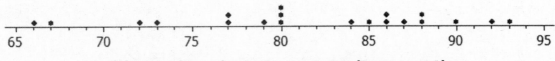

Number of French Fries in a Large Bag (Restaurant A)

Sometimes it is useful to know what point separates a data distribution into two equal parts, where one part represents the upper half of the data values and the other part represents the lower half of the data values. This point is called the *median*. When the data are arranged in order from smallest to largest, the same number of values will be above the median point as below the median.

Exercises 1–3

1. You just bought a large bag of fries from the restaurant. Do you think you have exactly 82 french fries? Why or why not?

2. How many bags were in the sample?

3. Which of the following statement(s) would seem to be true for the given data? Explain your reasoning.

    a. Half of the bags had more than 82 fries in them.

    b. Half of the bags had fewer than 82 fries in them.

    c. More than half of the bags had more than 82 fries in them.

    d. More than half of the bags had fewer than 82 fries in them.

    e. If you got a random bag of fries, you could get as many as 93 fries.

Example 2

Examine the dot plot below.

Grades on a Science Test

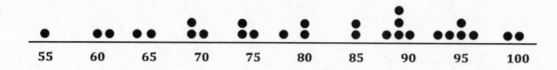

    a. How many data values are represented on the dot plot above?

    b. How many data values should be located above the median? How many below the median? Explain.

    c. For this data set, 14 values are 80 or smaller, and 14 values are 85 or larger, so the median should be between 80 and 85. When the median falls between two values in a data set, we use the average of the two middle values. For this example, the two middle values are 80 and 85. What is the median of the data presented on the dot plot?

EUREKA
MATH®

    d.   What does this information tell us about the data?

Example 3

Use the information from the dot plot in Example 2.

    a.   What percentage of students scored higher than the median? Lower than the median?

    b.   Suppose the teacher made a mistake, and the student who scored 65 actually scored a 71. Would the median change? Why or why not?

    c.   Suppose the student who scored a 65 actually scored an 89. Would the median change? Why or why not?

Example 4

A grocery store usually has three checkout lines open on Saturday afternoons. One Saturday afternoon, the store manager decided to count how many customers were waiting to check out at 10 different times. She calculated the median of her ten data values to be 8 customers.

    a.   Why might the median be an important number for the store manager to consider?

b. Give another example of when the median of a data set might provide useful information. Explain your thinking.

Exercises 4–5: A Skewed Distribution

4. The owner of the chain decided to check the number of french fries at another restaurant in the chain. Here are the data for Restaurant B: 82, 83, 83, 79, 85, 82, 78, 76, 76, 75, 78, 74, 70, 60, 82, 82, 83, 83, 83.

a. How many bags of fries were counted?

b. Sallee claims the median is 75 because she sees that 75 is the middle number in the data set listed above. She thinks half of the bags had fewer than 75 fries because there are 9 data values that come before 75 in the list, and there are 9 data values that come after 75 in the list. Do you think she would change her mind if the data were plotted in a dot plot? Why or why not?

c. Jake said the median was 83. What would you say to Jake?

d. Betse argued that the median was halfway between 60 and 85, or 72.5. Do you think she is right? Why or why not?

Lesson 12:   Describing the Center of a Distribution Using the Median

EUREKA
MATH

e.   Chris thought the median was 82. Do you agree? Why or why not?

5.   Calculate the mean, and compare it to the median. What do you observe about the two values? If the mean and median are both measures of center, why do you think one of them is smaller than the other?

Exercises 6–8: Finding Medians from Frequency Tables

6.   A third restaurant (Restaurant C) tallied the number of fries for a sample of bags of french fries and found the results below.

Number of Fries	Frequency
75	II
76	I
77	II
78	III
79	++++
80	IIII
81	I
82	I
83	
84	III
85	III
86	I

a.   How many bags of fries did they count?

b. What is the median number of fries for the sample of bags from this restaurant? Describe how you found your answer.

7. Robere wanted to look more closely at the data for bags of fries that contained a smaller number of fries and bags that contained a larger number of fries. He decided to divide the data into two parts. He first found the median of the whole data set and then divided the data set into the bottom half (the values in the ordered list that are before the median) and the top half (the values in the ordered list that are after the median).

a. List the 13 values in the bottom half. Find the median of these 13 values.

b. List the 13 values of the top half. Find the median of these 13 values.

8. Which of the three restaurants seems most likely to really have 82 fries in a typical bag? Explain your thinking.

Lesson 12:   Describing the Center of a Distribution Using the
              Median

Lesson Summary

The **median** is the middle value (or the mean of the two middle values) in a data set that has been ordered from smallest to largest. The median separates the data into two parts with the same number of data values below the median as above the median in the ordered list. To find a median, you first have to order the data. For an even number of data values, you find the average of the two middle numbers. For an odd number of data values, you use the middle value.

Name _____   Date _____

1. What is the median age for the following data set representing the ages of students requesting tickets for a summer band concert? Explain your reasoning.

13 14 15 15 16 16 17 18 18

2. What is the median number of diseased trees from a data set representing the numbers of diseased trees on each of 12 city blocks? Explain your reasoning.

11 3 3 4 6 12 9 3 8 8 8 1

3. Describe how you would find the median for a set of data that has 35 values. How would this be different if there were 36 values?

1. Make up a data set such that the following is true:

    a. The data set has 15 different values, and the median is 9.

    Possible Set: 0, 1, 2, 3, 4, 5, 8, 9, 10, 13, 14, 15, 18, 19, 20

    > I know that because this question asks for 15 different values, none of the numbers can repeat. Since 9 is the median, there should be 7 values that are lower than 9 and 7 values greater than 9, leaving 9 in the middle when ordered from least to greatest.

    b. The data set has 8 values, and the median is 32.

    Possible Set: 10, 14, 18, 30, 34, 35, 38, 40

    > In this question, I can repeat numbers. There must be an even number of values, so that means that two of the numbers will be in the middle. I could place two 32's in the middle. Or, I could choose two numbers with a mean of 32.

    c. The data set has 5 values, and the median is the same as the greatest value.

    Possible Set: 3, 5, 11, 11, 11

    > When I find the median, I start by putting the numbers in order from least to greatest. If they are in correct order and the median and the greatest are the same, then any numbers between the two will also have to be the same.

2. The dot plot shows the number of miles each person in a random sample ran last week.

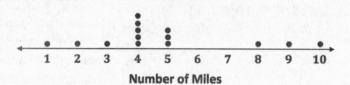

Weekly Running Totals

Number of Miles

a. How many people were in the sample?

    There are 14 people in the sample.

    > Each one of the dots represents a response from a person, so I can just count the dots to see how many people were in the sample.

b. Find the median number of miles run in the sample.

    1, 2, 3, 4, 4, 4, 4, 4, 5, 5, 5, 8, 9, 10

    The median number of miles run this week is 4.

    > If I have trouble using the dot plot to determine the median, I could write out all the numbers in order.

c. Do you think the mean or median would be a better description of the typical number of miles run? Explain your thinking.

    The mean is approximately 4.86 miles per week, while the median is 4 miles per week. The mean is slightly higher than the median and has a value that is greater than most of the dots on the dot plot. Therefore, the median is a closer representation.

    > Calculating the mean is a necessary step in answering this question.

Lesson 12:   Describing the Center of a Distribution Using the
                                Median

© 2019 Great Minds®. eureka-math.org

EUREKA
MATH

3. The salaries of eleven employees at a local business are given below.

Employee	Salary
President	$320,000
Vice President	$232,000
Manager	$94,000
Employee A	$64,000
Employee B	$64,000
Employee C	$58,000
Employee D	$51,000
Employee E	$50,000
Employee F	$48,000
Employee G	$48,000
Employee H	$47,000

> I notice that these numbers are already in order, so I just need to determine which number is in the middle to determine the median.

a. Find the median salary, and explain what it tells you about the salaries.

   The median salary is $58,000 *for Employee C. Half of the employees make more than* $58,000, *and half of the employees make less than* $58,000.

b. Find the median of the lower half of the salaries and the median of the upper half of the salaries.

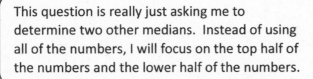

> This question is really just asking me to determine two other medians. Instead of using all of the numbers, I will focus on the top half of the numbers and the lower half of the numbers.

   $48,000 *is the median for the bottom half of the salaries.* $94,000 *is the median for the top half of the salaries.*

c. Find the width of each of the following intervals. What do you notice about the size of the interval widths, and what does that tell you about the salaries?

> I can use my answer to parts (a) and (b) to answer these questions. To get the size of the interval, I just need to find the difference between the values given.

i. Minimum salary to the median of the lower half: $\$1,000$

ii. Median of the lower half to the median of the whole data set: $\$10,000$

iii. Median of the whole data set to the median of the upper half: $\$36,000$

iv. Median of the upper half to the highest salary: $\$226,000$

The largest width is from the median of the upper half to the highest salary. The smaller salaries are closer together than the larger ones.

Lesson 12: Describing the Center of a Distribution Using the Median

EUREKA MATH

1. The amount of precipitation in each of the western states in the United States is given in the table as well as the dot plot.

State	Amount of Precipitation (inches)
WA	38.4
OR	27.4
CA	22.2
MT	15.3
ID	18.9
WY	12.9
NV	9.5
UT	12.2
CO	15.9
AZ	13.6
NM	14.6
AK	58.3
HI	63.7

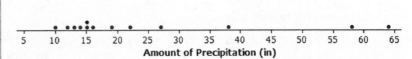

Amount of Precipitation (in)

Source: http://www.currentresults.com/Weather/US/average-annual-state-precipitation.php

a. How do the amounts vary across the states?

b. Find the median. What does the median tell you about the amount of precipitation?

c. Do you think the mean or median would be a better description of the typical amount of precipitation? Explain your thinking.

2. Identify the following as true or false. If a statement is false, give an example showing why.

a. The median is always equal to one of the values in the data set.

b. The median is halfway between the least and greatest values in the data set.

c. At most, half of the values in a data set have values less than the median.

d. In a data set with 25 different values, if you change the two smallest values in the data set to smaller values, the median will not be changed.

e. If you add 10 to every value in a data set, the median will not change.

3. Make up a data set such that the following is true:

a. The data set has 11 different values, and the median is 5.

b. The data set has 10 values, and the median is 25.

c. The data set has 7 values, and the median is the same as the least value.

4. The dot plot shows the number of landline phones that a sample of people have in their homes.

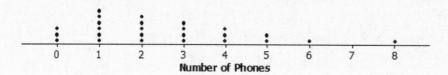

Number of Phones

a. How many people were in the sample?

b. Why do you think three people have no landline phones in their homes?

c. Find the median number of phones for the people in the sample.

5. The salaries of the Los Angeles Lakers for the 2012–2013 basketball season are given below. The salaries in the table are ordered from largest to smallest.

Player	Salary
Kobe Bryant	$27,849,149
Dwight Howard	$19,536,360
Pau Gasol	$19,000,000
Steve Nash	$8,700,000
Metta World Peace	$7,258,960
Steve Blake	$4,000,000
Jordan Hill	$3,563,600
Chris Duhon	$3,500,000
Jodie Meeks	$1,500,000
Earl Clark	$1,240,000
Devin Ebanks	$1,054,389
Darius Morris	$962,195
Antawn Jamison	$854,389
Robert Sacre	$473,604
Darius Johnson-Odom	$203,371

Source: www.basketball-reference.com/contracts/LAL.html

a. Just looking at the data, what do you notice about the salaries?

b. Find the median salary, and explain what it tells you about the salaries.

c. Find the median of the lower half of the salaries and the median of the upper half of the salaries.

d. Find the width of each of the following intervals. What do you notice about the size of the interval widths, and what does that tell you about the salaries?

i. Minimum salary to the median of the lower half:

ii. Median of the lower half to the median of the whole data set:

iii. Median of the whole data set to the median of the upper half:

iv. Median of the upper half to the highest salary:

6. Use the salary table from above to answer the following.

   a. If you were to find the mean salary, how do you think it would compare to the median? Explain your reasoning.

   b. Which measure do you think would give a better picture of a typical salary for the Lakers, the mean or the median? Explain your thinking.

In Lesson 12, the median was used to describe a typical value for a data set. But the values in a data set vary around the median. What is a good way to indicate how the data vary when we use a median as an indication of a typical value? These questions are explored in the following exercises.

Exercises 1–4: More French Fries

1. In Lesson 12, you thought about the claim made by a chain restaurant that the typical number of french fries in a large bag was 82. Then, you looked at data on the number of fries in a bag from three of the restaurants.

   a. How do you think the data were collected, and what problems might have come up in collecting the data?

   b. What scenario(s) would give counts that might not be representative of typical bags?

2. The medians of the top half and the medians of the bottom half of the data for each of the three restaurants are as follows: Restaurant A—87.5 and 77; Restaurant B—83 and 76; Restaurant C—84 and 78. The difference between the medians of the two halves is called the *interquartile range*, or IQR.

   a. What is the IQR for each of the three restaurants?

   b. Which of the restaurants had the smallest IQR, and what does that tell you?

c. The median of the bottom half of the data is called the *lower quartile* (denoted by Q1), and the median of the top half of the data is called the *upper quartile* (denoted by Q3). About what fraction of the data would be between the lower and upper quartiles? Explain your thinking.

3. Why do you think that the median of the top half of the data is called the *upper quartile* and the median of the bottom half of the data is called the *lower quartile*?

4.

a. Mark the quartiles for each restaurant on the graphs below.

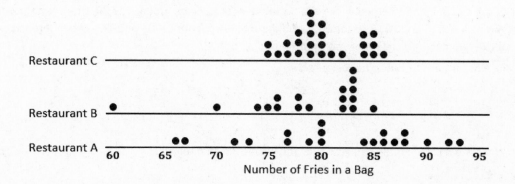

b. Does the IQR help you decide which of the three restaurants seems most likely to really have 82 fries in a typical large bag? Explain your thinking.

Describing Variability Using the Interquartile Range (IQR)

EUREKA MATH

Example 1: Finding the IQR

Read through the following steps. If something does not make sense to you, make a note, and raise it during class discussion. Consider the data: 1, 1, 3, 4, 6, 6, 7, 8, 10, 11, 11, 12, 15, 15, 17, 17, 17

Creating an IQR:

a. Put the data in order from smallest to largest.

b. Find the minimum and maximum.

c. Find the median.

d. Find the lower quartile and upper quartile.

e. Calculate the IQR by finding the difference between Q3 and Q1.

Exercise 5: When Should You Use the IQR?

5. When should you use the IQR? The data for the 2012 salaries for the Lakers basketball team are given in the two plots below. (See Problem 5 in the Problem Set from Lesson 12.)

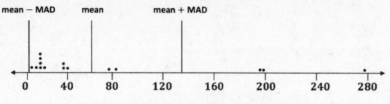

Salary (hundreds of thousands of dollars)

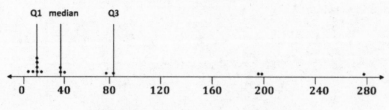

Salary (hundreds of thousands of dollars)

a. The data are given in hundreds of thousands of dollars. What would a salary of 40 hundred thousand dollars be?

b. The vertical lines on the top plot show the mean and the mean plus and minus the MAD. The bottom plot shows the median and the IQR. Which interval is a better picture of the typical salaries? Explain your thinking.

EUREKA
MATH·

Exercise 6: On Your Own with IQRs

6. Create three different examples where you might collect data and where that data might have an IQR of 20. Define a median in the context of each example. Be specific about how the data might have been collected and the units involved. Be ready to describe what the median and IQR mean in each context.

   a.

   b.

   c.

Lesson Summary

To find the IQR, you order the data, find the median of the data, and then find the median of the bottom half of the data (the lower quartile) and the median of the top half of the data (the upper quartile). The IQR is the difference between the upper quartile and the lower quartile, which is the length of the interval that includes the middle half of the data. The median and the two quartiles divide the data into four sections, with about $\frac{1}{4}$ of the data in each section. Two of the sections are between the quartiles, so the interval between the quartiles would contain about 50% of the data.

Name _____   Date _____

1. On the dot plot below, insert the following words in approximately the correct position.

Maximum   Minimum   IQR   Median   Lower Quartile (Q1)   Upper Quartile (Q3)

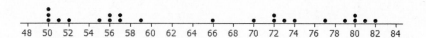

2. Estimate the IQR for the data set shown in the dot plot.

1. In each of parts (a)–(b), create a data set with at least 6 values such that it has the following properties:

   a. An IQR equal to 12.

   One example is {3, 8, 11, 13, 14, 20, 23}
   where the IQR is 12 because 20 − 8 = 12.

   > For the IQR to be equal to 12, $Q3 − Q1$ must also be equal to 12. So Q3 could be 20, and Q1 could be 8, making a difference of 12.

   b. An IQR equal to 0.

   One example is {10, 18, 18, 18, 18, 18, 24}.

   > For the IQR to be equal to 0, Q3 and Q1 must be the same. If the median of the lower half of the data is the same as the median of the top half of the data, all the values between the two must also be the same.

2. A sample of the heights of students in two classes are given, in inches, in the table below.

Mrs. M's Class	44	38	47	46	39	42	40	46	35	46
Mr. V's Class	52	58	42	38	45	40	62	56	45	49

   a. How do you think the data might have been collected?

   Someone at the school may have measured the students' heights. The measuring may have been
   done by the teacher or the school nurse.

b.   Do you think it would be possible for $\frac{1}{4}$ of the heights for Mr. V's class to be 50 inches or above? Why or why not?

Yes, it might be possible. The mean height in Mr. V's class is 48.7 inches, and more than 25% of the sample values are 50 inches or above.

c.   Make a prediction about how the values of the IQR for the heights for each class compare. Explain your thinking.

Mr. V's class probably has the larger IQR because those heights seem to vary more than the heights for Mrs. M's class.

I know that the IQR measures variability. I can look at the data and see which set varies more in order to make my prediction.

d.   Find the IQR for the heights for each class. How do the results compare to what you predicted?

Mrs. M's class: 35, 38, 39, 40, 42, 44, 46, 46, 46, 47

$Q1 = 39$, **Median** $= 43$, $Q3 = 46$

In order to calculate the IQR for each class, I must first place the heights in order from least to greatest. Then, I can determine the key points, like the median, Q1, and Q3.

Mr. V's class: 38, 40, 42, 45, 45, 49, 52, 56, 58, 62

$Q1 = 42$, **Median** $= 47$, $Q3 = 56$

The IQR for Mrs. M's class is 7 inches because $46 - 39 = 7$. For Mr. V's class, the IQR is 14 inches because $56 - 42 = 14$. This result matches my prediction in part (c).

       Lesson 13:   Describing Variability Using the Interquartile Range (IQR)

EUREKA MATH

1. The average monthly high temperatures (in degrees Fahrenheit) for St. Louis and San Francisco are given in the table below.

	Jan.	Feb.	Mar.	Apr.	May	June	July	Aug.	Sept.	Oct.	Nov.	Dec.
St. Louis	40	45	55	67	77	85	89	88	81	69	56	43
San Francisco	57	60	62	63	64	67	67	68	70	69	63	57

Data Source: http://www.weather.com

a. How do you think the data might have been collected?

b. Do you think it would be possible for $\frac{1}{4}$ of the temperatures in the month of July for St. Louis to be 95°F or above? Why or why not?

c. Make a prediction about how the values of the IQR for the temperatures for each city compare. Explain your thinking.

d. Find the IQR for the average monthly high temperature for each city. How do the results compare to what you predicted?

2. The plot below shows the years in which each of 100 pennies were made.

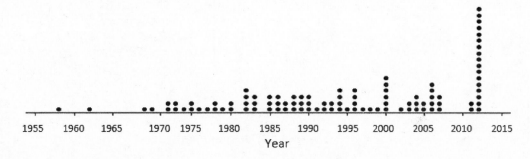

Year

a. What does the stack of 17 dots at 2012 representing 17 pennies tell you about the age of these pennies in 2014?

b. Here is some information about the sample of 100 pennies. The mean year they were made is 1994; the first year any of the pennies were made was 1958; the newest pennies were made in 2012; Q1 is 1984, the median is 1994, and Q3 is 2006; the MAD is 11.5 years. Use the information to indicate the years in which the middle half of the pennies was made.

3. In each of parts (a)–(c), create a data set with at least 6 values such that it has the following properties:

a. A small IQR and a big range (maximum − minimum)

b. An IQR equal to the range

c. The lower quartile is the same as the median.

4. Rank the following three data sets by the value of the IQR.

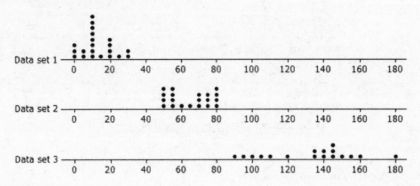

5. Here are the number of fries in each of the bags from Restaurant A:

    $$80, 72, 77, 80, 90, 85, 93, 79, 84, 73, 87, 67, 80, 86, 92, 88, 86, 88, 66, 77$$

    a. Suppose one bag of fries had been overlooked and that bag had only 50 fries. If that value is added to the data set, would the IQR change? Explain your reasoning.

    b. Will adding another data value always change the IQR? Give an example to support your answer.

EUREKA
MATH

A box plot is a graph that is used to summarize a data distribution. What does the box plot tell us about the data distribution? How does the box plot indicate the variability of the data distribution? These questions are explored in this lesson.

Example 1: Time to Get to School

Consider the statistical question, "What is the typical amount of time it takes for a person in your class to get to school?" The amount of time it takes to get to school in the morning varies for the students in your class. Take a minute to answer the following questions. Your class will use this information to create a dot plot.

Write your name and an estimate of the number of minutes it took you to get to school today on a sticky note.

What were some of the things you had to think about when you made your estimate?

Exercises 1–4

Here is a dot plot of the estimates of the times it took students in Mr. S's class to get to school one morning.

Mr. S's Class

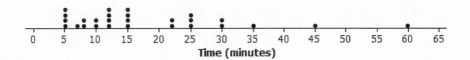

Time (minutes)

1. Put a line on the dot plot that you think separates the times into two groups—one group representing the longer times and the other group representing the shorter times.

EUREKA
MATH®

2. Put another line on the dot plot that separates out the times for students who live really close to the school. Add another line that separates out the times for students who take a very long time to get to school.

3. Your dot plot should now be divided into four sections. Record the number of data values in each of the four sections.

4. Share your marked-up dot plot with some of your classmates. Compare how each of you divided the dot plot into four sections.

Exercises 5–7: Time to Get to School

The times (in minutes) for the students in Mr. S's class have been put in order from smallest to largest and are shown below.

5 5 5 5 7 8 8 10 10 12 12 12 12 15 15 15 15 22 22 25 25 25 30 30 35 45 60

5. What is the value of the median time to get to school for students in Mr. S's class?

6. What is the value of the lower quartile? The upper quartile?

EUREKA
MATH®

7. The lines on the dot plot below indicate the location of the median, the lower quartile, and the upper quartile. These lines divide the data set into four parts. About what fraction of the data values are in each part?

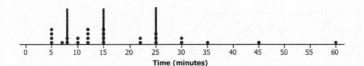

Mr. S's Class

Example 2: Making a Box Plot

A box plot is a graph made using the following five numbers: the smallest value in the data set, the lower quartile, the median, the upper quartile, and the largest value in the data set.

To make a box plot:

- Find the median of all of the data.
- Find Q1, the median of the bottom half of the data, and Q3, the median of the top half of the data.
- Draw a number line, and then draw a box that goes from Q1 to Q3.
- Draw a vertical line in the box at the value of the median.
- Draw a line segment connecting the minimum value to the box and a line segment that connects the maximum value to the box.

You will end up with a graph that looks something like this:

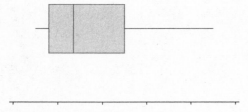

Now, use the given number line to make a box plot of the data below.

20, 21, 25, 31, 35, 38, 40, 42, 44

```
|_____|
   15      20      25      30      35      40      45
```

The five-number
summary is as follows:

Min =

Q1 =

Median =

Q3 =

Max =

Exercises 8–11: A Human Box Plot

Consider again the sticky note that you used to write down the number of minutes it takes you to get to school. If possible, you and your classmates will form a human box plot of the number of minutes it takes students in your class to get to school.

8. Find the median of the group. Does someone represent the median? If not, who is the closest to the median?

9. Find the maximum and minimum of the group. Who are they?

10. Find Q1 and Q3 of the group. Does anyone represent Q1 or Q3? If not, who is the closest to Q1? Who is the closest to Q3?

11. Sketch the box plot for this data set.

EUREKA
MATH

Lesson Summary

You learned how to make a box plot by doing the following:

- Finding the median of the entire data set.

- Finding Q1, the median of the bottom half of the data, and Q3, the median of the top half of the data.

- Drawing a number line and then drawing a box that goes from Q1 to Q3.

- Drawing a vertical line in the box at the value of the median.

- Drawing a line segment connecting the minimum value to the box and one that connects the maximum value to the box.

Name _____   Date _____

Sulee explained how to make a box plot to her sister as follows:

"First, you find the smallest and largest values and put a mark halfway between them, and then put a mark halfway between that mark and each end. So, if 10 is the smallest value and 30 is the largest value, you would put a mark at 20. Then, another mark belongs halfway between 20 and 10, which would be at 15. And then one more mark belongs halfway between 20 and 30, which would be at 25. Now, you put a box around the three middle marks, and draw lines from the box to the smallest and largest values."

Here is her box plot. What would you say to Sulee?

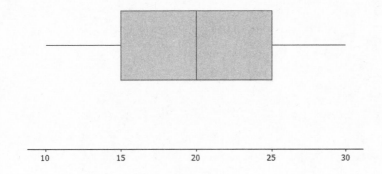

1. The box plot below summarizes data from a survey of households about the number of televisions they have. Identify each of the following statements as true or false. Explain your reasoning in each case.

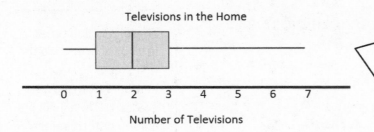

Televisions in the Home

Number of Televisions

> I know that a box plot is made using 5 key points: the lowest number, the median of the lower half of the numbers, the median of the entire set, the median of the upper half of the numbers, and the greatest number. The lines and box show me these five values from left to right. This makes up the five-number summary.

a. The maximum number of televisions per house is 3.

   False. The line segment at the top reaches 7.

b. At least $\frac{1}{2}$ of the houses have 1 or more televisions.

   True. 2 is the median. This tells us that half the houses have 2 or more televisions. If we use the lower median, we can see that three-fourths of the houses have at least one television, which is more than one-half.

c. All of the houses have televisions.

   False. The lower line segment starts at 0, so at least one household does not have a television.

d. Half of the houses surveyed have between 0 and 2 televisions.

   True. About 25% of the houses would have between 0 and 1 television, and another 25% of the houses would have between 1 and 2 televisions, making a total of 50%, or half, of the houses.

2. The number of minutes it takes a group of runners to complete a 5K race are as follows:

    12, 18, 24, 30, 45, 22, 18, 42, 32, 38, 28, 28, 28, 24, 25, 16, 39, 21

    a. Make a box plot of the finishing times.

    12, 16, 18, 18, 21, 22, 24, 24, 25, 28, 28, 28, 30, 32, 38, 39, 42, 45

    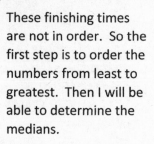

    These finishing times are not in order. So the first step is to order the numbers from least to greatest. Then I will be able to determine the medians.

    Five-number summary: **12, 21, 26. 5, 32, 45**

    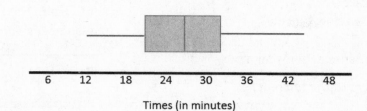

    5K Finishing Times

    Times (in minutes)

    b. Describe the finishing times distribution. Include a description of center and spread.

    The IQR is 32 minutes − 21 minutes, or 11 minutes. Half of the finishing times were near the middle between 21 minutes and 32 minutes. The median is 26. 5 minutes. A quarter of the finishing times are less than 21 minutes but greater than or equal to 12 minutes. A quarter of the finishing times are greater than 32 minutes and less than or equal to 45 minutes. The finishing times varied from 12 minutes to 45 minutes.

    I know that my box plot has four sections and that each section represents a quarter of the data values. I can use this to help me describe the distribution.

Lesson 14:   Summarizing a Distribution Using a Box Plot

EUREKA MATH

1. Dot plots for the amount of time it took students in Mr. S's and Ms. J's classes to get to school are below.

Mr. S's Class   **Ms. J's Class**

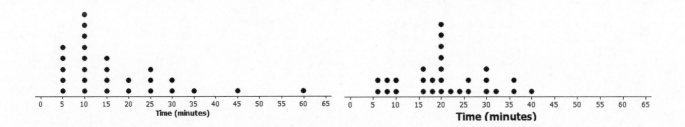

a. Make a box plot of the times for each class.

b. What is one thing you can see in the dot plot that you cannot see in the box plot? What is something that is easier to see in the box plot than in the dot plot?

2. The dot plot below shows the vertical jump of some NBA players. A vertical jump is how high a player can jump from a standstill. Draw a box plot of the heights for the vertical jumps of the NBA players above the dot plot.

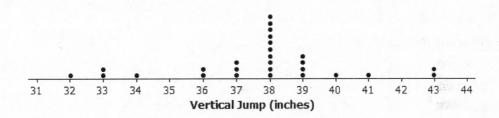

3. The mean daily temperatures in degrees Fahrenheit for the month of February for a certain city are as follows:

    4, 11, 14, 15, 17, 20, 30, 23, 20, 35, 35, 31, 34, 23, 15, 19, 39, 22, 15, 15, 19, 39, 22, 23, 29, 26, 29, 29

a. Make a box plot of the temperatures.

b. Make a prediction about the part of the United States you think the city might be located. Explain your reasoning.

c. Describe the temperature data distribution. Include a description of center and spread.

4. The box plot below summarizes data from a survey of households about the number of dogs they have. Identify each of the following statements as true or false. Explain your reasoning in each case.

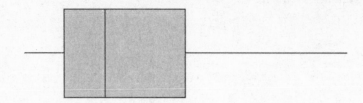

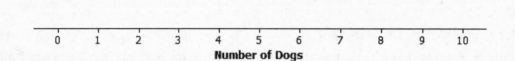

Number of Dogs

a. The maximum number of dogs per house is 8.

b. At least $\frac{1}{2}$ of the houses have 2 or more dogs.

c. All of the houses have dogs.

d. Half of the houses surveyed have between 2 and 4 dogs.

e. Most of the houses surveyed have no dogs.

EUREKA MATH

You reach into a jar of Tootsie Pops. How many Tootsie Pops do you think you could hold in one hand? Do you think the number you could hold is greater than or less than what other students can hold? Is the number you could hold a typical number of Tootsie Pops? This lesson examines these questions.

Example 1: Tootsie Pops

Ninety-four people were asked to grab as many Tootsie Pops as they could hold. Here is a box plot for these data. Are you surprised?

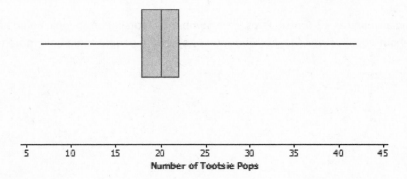

Exercises 1–5

1. What might explain the variability in the number of Tootsie Pops that the 94 people were able to hold?

2. Use a box plot to estimate the values in the five-number summary.

3. Describe how the box plot can help you understand differences in the numbers of Tootsie Pops people could hold.

4. Here is Jayne's description of what she sees in the box plot. Do you agree or disagree with her description? Explain your reasoning.

"One person could hold as many as 42 Tootsie Pops. The number of Tootsie Pops people could hold was really different and spread about equally from 7 to 42. About one-half of the people could hold more than 20 Tootsie Pops."

5. Here is a different box plot of the same data on the number of Tootsie Pops 94 people could hold.

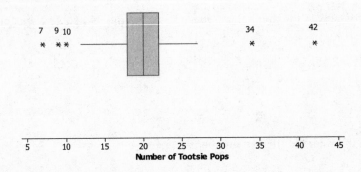

a. Why do you suppose there are five values that are shown as separate points and are labeled?

b. Does knowing these data values change anything about your responses to Exercises 1 to 4 above?

EUREKA MATH

Exercises 6–10: Maximum Speeds

The maximum speeds of selected birds and land animals are given in the tables below.

Bird	Speed (mph)
Peregrine falcon	242
Swift bird	120
Spine-tailed swift	106
White-throated needle tail	105
Eurasian hobby	100
Pigeon	100
Frigate bird	95
Spur-winged goose	88
Red-breasted merganser	80
Canvasback duck	72
Anna's hummingbird	61.06
Ostrich	60

Land Animal	Speed (mph)
Cheetah	75
Free-tailed bat (in flight)	60
Pronghorn antelope	55
Lion	50
Wildebeest	50
Jackrabbit	44
African wild dog	44
Kangaroo	45
Horse	43.97
Thomson's gazelle	43
Greyhound	43
Coyote	40
Mule deer	35
Grizzly bear	30
Cat	30
Elephant	25
Pig	9

Data sources: *Natural history Magazine*, March 1974, copyright 1974; The American Museum of Natural History; and James G. Doherty, general curator, The Wildlife Conservation Society; http://www.thetravelalmanac.com/lists/animals-speed.htm; http://en.wikipedia.org/wiki/Fastest animals

6. As you look at the speeds, what strikes you as interesting?

7. Do birds or land animals seem to have the greatest variability in speeds? Explain your reasoning.

8. Find the five-number summary for the speeds in each data set. What do the five-number summaries tell you about the distribution of speeds for each data set?

9. Use the five-number summaries to make a box plot for each of the two data sets.

| 0 | 25 | 50 | 75 | 100 | 125 | 150 | 175 | 200 | 225 | 250 |

Maximum Speed of Birds (mph)

| 0 | 25 | 50 | 75 | 100 | 125 | 150 | 175 | 200 | 225 | 250 |

Maximum Speed of Land Animals (mph)

10. Write several sentences describing the speeds of birds and land animals.

EUREKA MATH

Exercises 11–15: What Is the Same, and What Is Different?

Consider the following box plots, which show the number of correctly answered questions on a 20-question quiz for students in three different classes.

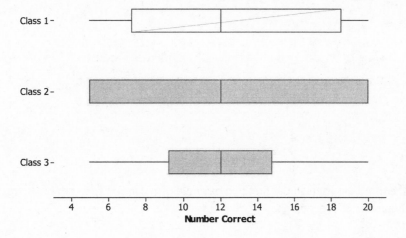

11. Describe the variability in the scores of each of the three classes.

12.

   a. Estimate the interquartile range for each of the three sets of scores.

   b. What fraction of students would have scores in the interval that extends from the lower quartile to the upper quartile?

   c. What does the value of the IQR tell you about how the scores are distributed?

13. Which class do you believe performed the best? Be sure to use information from the box plots to back up your answer.

14.

    a. Find the IQR for the three data sets in the first two examples: maximum speed of birds, maximum speed of land animals, and number of Tootsie Pops.

    b. Which data set had the highest percentage of data values between the lower quartile and the upper quartile? Explain your thinking.

15. A teacher asked students to draw a box plot with a minimum value at 34 and a maximum value at 64 that had an interquartile range of 10. Jeremy said he could not draw just one because he did not know where to put the box on the number line. Do you agree with Jeremy? Why or why not?

Lesson 15:   More Practice with Box Plots

EUREKA MATH

Name _____   Date _____

Given the following information, create a box plot, and find the IQR.

For a large group of dogs, the shortest dog was 6 inches, and the tallest was 32 inches. One-half of the dogs were taller than 18 inches. One-fourth of the dogs were shorter than 15 inches. The upper quartile of the dog heights was 23 inches.

```
 ——————————————————————————————————————————————
   4   6   8   10  12  14  16  18  20  22  24  26  28  30  32  34
                       Dog Height (inches)
```

1. The box plot below summarizes the average time students in Ms. Baker's math class spend on homework every night.

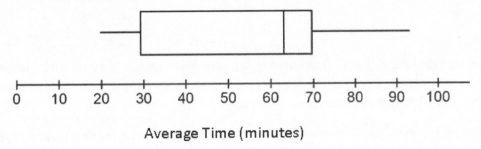

Average Time (minutes)

a. Estimate the values in the five-number summary from the box plot.

    Answers may vary. Minimum = 20 min; Q1 = 30 min; Median = 63 min; Q3 = 70 min; Maximum = 93 min

b. The highest average time is 93 minutes, followed by the second highest average time of 77 minutes. What does this tell you about the spread of the average times spent on math homework every night in the top quarter of the box plot?

    The Q3 is about 70 minutes, so all but one of the scores in the top quarter are between 70 minutes and 77 minutes.

> If approximately $\frac{1}{4}$ of the data values are between Q3 (70) and the maximum (93) and the second highest average is 77, then I know there are no more values between 77 and 93.

c. Use the five-number summary and the IQR to describe the average amount of time students spend on math homework every night.

$$IQR = Q3 - Q1$$

The average times vary from 20 minutes to 93 minutes. The IQR is 40 minutes; the middle half of the average times are between 30 minutes and 70 minutes. Half of the times students spent on homework are less than 63 minutes.

> The middle half of the data is between Q1 and Q3.

> The median is 63 and about one half of the data is above/below the median.

2. Suppose the interquartile range for the number of hours students spent playing outside during the summer was 10. What do you think about each of the following statements? Explain your reasoning.

a. About half of the students played outside for 10 hours during the summer.

This may not be correct as you know the width of the interval that contains the middle half of the times was 10, but you do not know where it starts or stops. You do not know the lower or upper quartile.

> I remember the interquartile range describes how spread out the middle 50% of the data are in the data distribution.

b. All of the students played at least 10 hours outside during the summer.

This may not be correct for the same reason as in part (a).

c. About half of the class could have played outside from 10 to 20 hours or from 15 to 25 hours.

Either could be correct as the only information given is the interquartile range of 10, and the statement says "could have."

> About half of the class could have also played from 9 to 19 hours or from 12 to 22 hours. There are many possibilities.

Lesson 15:   More Practice with Box Plots

EUREKA MATH

3. Suppose you know the following for a data set: The minimum value is 65, the lower quartile is 77, the IQR is 21, half of the data are less than 85, and the maximum value is 105.

   a. Think of a context for which these numbers might make sense.

      Answers will vary. One possibility is a healthy person's normal resting heart rate in beats per minute (bpm) since the resting heart rate for a healthy person is between 60 bpm and 100 bpm.

   b. Sketch a box plot related to the context in part (a).

   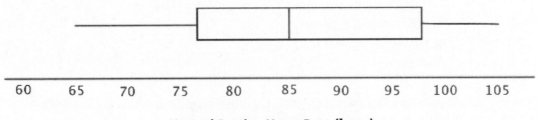

   | 60 | 65 | 70 | 75 | 80 | 85 | 90 | 95 | 100 | 105 |

   Normal Resting Heart Rate (bpm)

   > I need to determine the upper quartile to construct the box plot. I know the IQR is calculated by finding the difference of the upper quartile and the lower quartile. Since the IQR and the lower quartile are given, I can write an equation to find the value for the upper quartile.
   >
   > Upper Quartile − Lower Quartile = Interquartile Range
   >
   > Upper Quartile = Interquartile Range + Lower Quartile
   >
   > Upper Quartile = 21 + 77
   >
   > Upper Quartile = 98

   c. Are there more data values above or below this median? Explain your reasoning.

      The number of data values on either side of the median should be about the same, one half of all of the data.

4. The speeds of the fastest birds are given in the table below.

Type of Bird	Speed (mph)
Peregrine Falcon	242
Golden Eagle	199
Gyrfalcon	130
Common Swift	106
Eurasian Hobby	100
Frigatebird	95
Spur Winged Goose	88
Homing Pigeon	87
Red-breasted Merganser	81
White-rumped Swift	77
Canvasback	73

Data Source: http://dinoanimals.com/animals/the-fastest-birds-in-the-world-top-10/, accessed September 28, 2015

a. Find the five-number summary for this data set, and use it to create a box plot of the speeds.

Minimum = 73, Q1 = 81, Median = 95, Q3 = 130, Maximum = 242

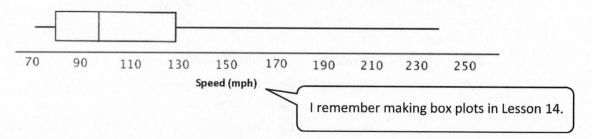

I remember making box plots in Lesson 14.

b. Why is the median not in the center of the box?

The median is not in the center of the box because about $\frac{1}{4}$ of the speeds are between 95 mph *and* 130 mph, *and another $\frac{1}{4}$ are closer together, between* 81 mph *and* 95 mph.

I remember approximately $\frac{1}{4}$ of the data values are found in each section of a box plot.

c. Write a few sentences telling your friend about the speeds of the fastest birds.

Half of the birds fly faster than 95 mph; *the fastest bird in the list is the Peregrine Falcon with a speed of* 242 mph. *The slowest bird in the list is the Canvasback with a speed of* 73 mph. *The middle* 50% *of the speeds are between* 81 mph *and* 130 mph.

EUREKA MATH

1. The box plot below summarizes the maximum speeds of certain kinds of fish.

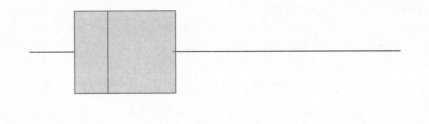

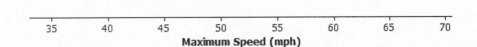

a. Estimate the values in the five-number summary from the box plot.

b. The fastest fish is the sailfish at 68 mph, followed by the marlin at 50 mph. What does this tell you about the spread of the fish speeds in the top quarter of the box plot?

c. Use the five-number summary and the IQR to describe the speeds of the fish.

2. Suppose the interquartile range for the number of hours students spent playing video games during the school week was 10. What do you think about each of the following statements? Explain your reasoning.

a. About half of the students played video games for 10 hours during a school week.

b. All of the students played at least 10 hours of video games during the school week.

c. About half of the class could have played video games from 10 to 20 hours a week or from 15 to 25 hours.

3. Suppose you know the following for a data set: The minimum value is 130, the lower quartile is 142, the IQR is 30, half of the data are less than 168, and the maximum value is 195.

a. Think of a context for which these numbers might make sense.

b. Sketch a box plot.

c. Are there more data values above or below the median? Explain your reasoning.

4. The speeds for the fastest dogs are given in the table below.

Breed	Speed (mph)
Greyhound	45
African wild dog	44
Saluki	43
Whippet	36
Basanji	35
German shepherd	32
Vizsla	32
Doberman pinscher	30

Breed	Speed (mph)
Irish wolfhound	30
Dalmatian	30
Border collie	30
Alaskan husky	28
Giant schnauzer	28
Jack Russell terrier	25
Australian cattle dog	20

Data source: http://www.vetstreet.com/our-pet-experts/meet-eight-of-the-fastest-dogs-on-the-planet; http://canidaepetfood.blogspot.com/2012/08/which-dog-breeds-are-fastest.html

a. Find the five-number summary for this data set, and use it to create a box plot of the speeds.

b. Why is the median not in the center of the box?

c. Write a few sentences telling your friend about the speeds of the fastest dogs.

EUREKA MATH

Exercise 1: Supreme Court Chief Justices

1.   The Supreme Court is the highest court of law in the United States, and it makes decisions that affect the whole country. The chief justice is appointed to the court and is a justice the rest of his life unless he resigns or becomes ill. Some people think that this means that the chief justice serves for a very long time. The first chief justice was appointed in 1789.

The table shows the years in office for each of the chief justices of the Supreme Court as of 2013:

Name	Number of Years	Year Appointed
John Jay	6	1789
John Rutledge	1	1795
Oliver Ellsworth	4	1796
John Marshall	34	1801
Roger Brooke Taney	28	1836
Salmon P. Chase	9	1864
Morrison R. Waite	14	1874
Melville W. Fuller	22	1888
Edward D. White	11	1910
William Howard Taft	9	1921
Charles Evens Hughes	11	1930
Harlan Fiske Stone	5	1941
Fred M. Vinson	7	1946
Earl Warren	16	1953
Warren E. Burger	17	1969
William H. Rehnquist	19	1986
John G. Roberts	8	2005

Data source: http://en.wikipedia.org/wiki/List of Justices of the Supreme Court of the United States

Use the table to answer the following:

a.   Which chief justice served the longest term, and which served the shortest term? How many years did each of these chief justices serve?

© 2019 Great Minds®. eureka-math.org

b. What is the median number of years these chief justices have served on the Supreme Court? Explain how you found the median and what it means in terms of the data.

c. Make a box plot of the years the justices served. Describe the shape of the distribution and how the median and IQR relate to the box plot.

d. Is the median halfway between the least and the most number of years served? Why or why not?

Exercises 2–3: Downloading Songs

2. A broadband company timed how long it took to download 232 four-minute songs on a dial-up connection. The dot plot below shows their results.

a. What can you observe about the download times from the dot plot?

b. Is it easy to tell whether or not 12.5 minutes is in the top quarter of the download times?

EUREKA
MATH

c.   The box plot of the data is shown below. Now, answer parts (a) and (b) above using the box plot.

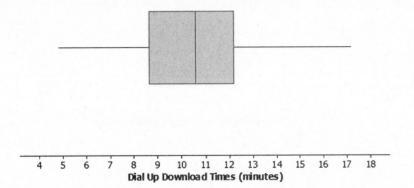

Dial Up Download Times (minutes)

d.   What are the advantages of using a box plot to summarize a large data set? What are the disadvantages?

3.   Molly presented the box plots below to argue that using a dial-up connection would be better than using a broadband connection. She argued that the dial-up connection seems to have less variability around the median even though the overall range seems to be about the same for the download times using broadband. What would you say?

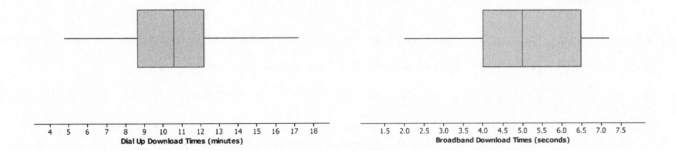

Dial Up Download Times (minutes)   Broadband Download Times (seconds)

Exercises 4–5: Rainfall

4. Data on the average rainfall for each of the twelve months of the year were used to construct the two dot plots below.

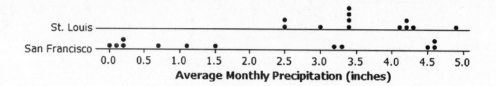

a. How many data points are in each dot plot? What does each data point represent?

b. Make a conjecture about which city has the most variability in the average monthly amount of precipitation and how this would be reflected in the IQRs for the data from both cities.

c. Based on the dot plots, what are the approximate values of the interquartile ranges (IQRs) for the average monthly precipitations for each city? Use the IQRs to compare the cities.

d. In an earlier lesson, the average monthly temperatures were rounded to the nearest degree Fahrenheit. Would it make sense to round the amount of precipitation to the nearest inch? Why or why not?

5. Use the data from Exercise 4 to answer the following.

    a. Make a box plot of the monthly precipitation amounts for each city using the same scale.

0.0	0.5	1.0	1.5	2.0	2.5	3.0	3.5	4.0	4.5	5.0

Average Monthly Precipitation in St. Louis (inches)

0.0	0.5	1.0	1.5	2.0	2.5	3.0	3.5	4.0	4.5	5.0

Average Monthly Precipitation in San Francisco (inches)

    b. Compare the percent of months that have above 2 inches of precipitation for the two cities. Explain your thinking.

    c. How does the top 25% of the average monthly precipitations compare for the two cities?

    d. Describe the intervals that contain the smallest 25% of the average monthly precipitation amounts for each city.

© 2019 Great Minds®. eureka-math.org

e. Think about the dot plots and the box plots. Which representation do you think helps you the most in understanding how the data vary?

Note: The data used in this problem are displayed in the table below.

Average Precipitation (inches)

	Jan.	Feb.	Mar.	Apr.	May	June	July	Aug.	Sept.	Oct.	Nov.	Dec.
St. Louis	2.45	2.48	3.36	4.10	4.80	4.34	4.19	3.41	3.38	3.43	4.22	2.96
San Francisco	4.5	4.61	3.76	1.46	0.70	0.16	0	0.06	0.21	1.12	3.16	4.56

Data source: http://www.weather.com

Name _____   Date _____

Data on the number of pets per family for students in a sixth-grade class are summarized in the box plot below:

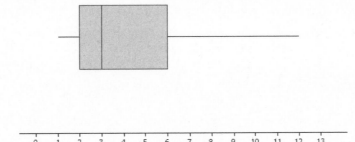

1. Can you tell how many families have two pets? Explain why or why not.

2. Given the box plot above, which of the following statements are true? If the statement is false, modify it to make the statement true.

    a. Every family has at least one pet.

    b. About one-fourth of the families have six or more pets.

    c. Most of the families have three pets.

    d. About half of the families have two or fewer pets.

    e. About three-fourths of the families have two or more pets.

The results of a jump rope competition between sixth and seventh graders are summarized below. Students recorded how many times they could jump rope in one minute.

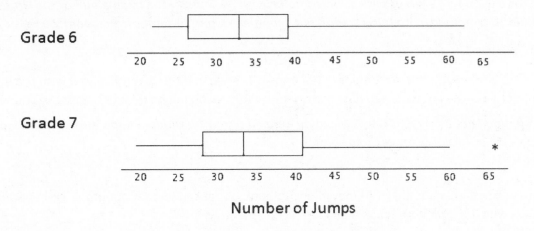

Number of Jumps

a. In which grade did the students do the best? Explain how you can tell.

Students were equally successful in both grades. For both grades, the median, the lower quartile and the upper quartile are about the same although these values for seventh grade are slightly shifted to the right.

I notice that the distribution of the data values in each set of data is very similar.

b. Why do you think one of the data values in Grade 7 is not part of the line segment?

The highest number of jumps was pretty far away from the other number of jumps, so it was marked separately.

c. How do the median number of jumps for the two grades compare? Is this surprising? Why or why not?

The median number of jumps in Grade 7 was about the same, but slightly higher, than the median number of jumps in Grade 6. This makes sense because the sixth and seventh graders should be able to jump approximately the same number of times in one minute. I wouldn't expect the seventh graders to jump many more times in one minute than the sixth graders, so these results did not surprise me.

d. How do the IQRs compare for the two grades?

The middle half of the Grade 7 number of jumps was fairly spread out spanning about 13 jumps from about 28 to 41 jumps with the median around 34 jumps. The middle half of the Grade 6 number of jumps was also fairly spread out spanning about 13 jumps from about 26 to 39 jumps.

e. The sixth grader with the most number of jumps was Max, and the seventh grader with the most number of jumps was Makayla. How many jumps did they do in one minute? How can you tell?

Max jumped about 63 times, and Makayla jumped about 66 times. I know this because I looked at the maximum value for each box plot.

f. Clara, a sixth grader, jumped 47 times in one minute. What can you say about the percent of sixth graders who jumped more times than Clara in one minute?

Clara is in the upper quartile with 47 jumps, so fewer than 25% of the sixth graders were able to jump more times than Clara.

1. The box plots below summarize the ages at the time of the award for leading actress and leading actor Academy Award winners.

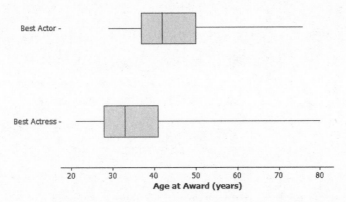

Data source: http://en.wikipedia.org/wiki/List_of_Best_Actor_winners_by_age_at_win

http://en.wikipedia.org/wiki/List_of_Best_Actress_winners_by_age_at_win

a. Based on the box plots, do you think it is harder for an older woman to win an Academy Award for best actress than it is for an older man to win a best actor award? Why or why not?

b. The oldest female to win an Academy Award was Jessica Tandy in 1990 for *Driving Miss Daisy*. The oldest actor was Henry Fonda for *On Golden Pond* in 1982. How old were they when they won the award? How can you tell? Were they a lot older than most of the other winners?

c. The 2013 winning actor was Daniel Day-Lewis for *Lincoln*. He was 55 years old at that time. What can you say about the percent of male award winners who were older than Daniel Day-Lewis when they won their Oscars?

d. Use the information provided by the box plots to write a paragraph supporting or refuting the claim that fewer older actresses than actors win Academy Awards.

2. The scores of sixth and seventh graders on a test about polygons and their characteristics are summarized in the box plots below.

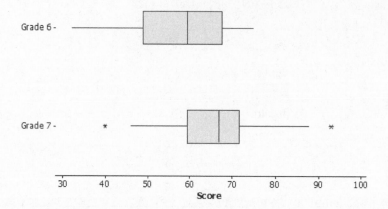

a. In which grade did the students do the best? Explain how you can tell.

b. Why do you think two of the data values for Grade 7 are not part of the line segments?

c. How do the median scores for the two grades compare? Is this surprising? Why or why not?

d. How do the IQRs compare for the two grades?

3. A formula for the IQR could be written as $Q3 - Q1 = IQR$. Suppose you knew the IQR and the Q1. How could you find the Q3?

4. Consider the statement, "Historically, the average length of service as chief justice on the Supreme Court has been less than 15 years; however, since 1969 the average length of service has increased." Use the data given in Exercise 1 to answer the following questions.

a. Do you agree or disagree with the statement? Explain your thinking.

b. Would your answer change if you used the median number of years rather than the mean?

EUREKA
MATH

Exploratory Challenge

Review of Statistical Questions

Statistical questions you investigated in this module included the following:

- How many hours of sleep do sixth graders typically get on a night when there is school the next day?
- What is the typical number of books read over the course of 6 months by a sixth grader?
- What is the typical heart rate of a student in a sixth-grade class?
- How many hours does a sixth grader typically spend playing a sport or a game outdoors?
- What are the head circumferences of adults interested in buying baseball hats?
- How long is the battery life of a certain brand of batteries?
- How many pets do students have?
- How long does it take students to get to school?
- What is a typical daily temperature in New York City?
- What is the typical weight of a backpack for students at a certain school?
- What is the typical number of french fries in a large order from a fast food restaurant?
- What is the typical number of minutes a student spends on homework each day?
- What is the typical height of a vertical jump for a player in the NBA?

What do these questions have in common?

Why do several of these questions include the word *typical*?

A Review of a Statistical Investigation

Recall from the very first lesson in this module that a statistical question is a question answered by data that you anticipate will vary.

Let's review the steps of a statistical investigation.

Step 1:   Pose a question that can be answered by data.

Step 2:   Collect appropriate data.

Step 3:   Summarize the data with graphs and numerical summaries.

Step 4:   Answer the question posed in Step 1 using the numerical summaries and graphs.

The first step is to pose a statistical question. Select one of the questions investigated in this module, and write it in the following Statistical Study Review Template.

The second step is to collect the data. In all of these investigations, you were given data. How do you think the data for the question you selected in Step 1 were collected? Write your answer in the summary below for Step 2.

The third step involves the various ways you summarize the data. List the various ways you summarized the data in the space for Step 3.

Statistical Study Review Template

Step 1: Pose a statistical question.
Step 2: Collect the data.
Step 3: Summarize the data.

Step 4: Answer the question.

Developing Statistical Questions

Now it is your turn to answer a statistical question based on data you collect. Before you collect the data, explore possible statistical questions. For each question, indicate the data that you would collect and summarize to answer the question. Also, indicate how you plan to collect the data.

Think of questions that could be answered by data collected from members of your class or school or data that could be collected from recognized websites (such as the American Statistical Association and the Census at School project). Your teacher will need to approve both your question and your plan to collect data before data are collected.

As a class, explore possibilities for a statistical investigation. Record some of the ideas discussed by your class using the following table.

Possible Statistical Questions	What data would be collected, and how would the data be collected?

© 2019 Great Minds®. eureka-math.org

After discussing several of the possibilities for a statistical project, prepare a statistical question and a plan to collect the data. After your teacher approves your question and data collection plan, begin collecting the data. Carefully organize your data as you begin developing the numerical and graphical summaries to answer your statistical question. In future lessons, you will be directed to begin creating a poster or an outline of a presentation that will be shared with your teacher and other members of your class.

Complete the following to present to your teacher:

1. The statistical question for my investigation is:

2. Here is how I propose to collect my data. (Include how you are going to collect your data and a clear description of what you plan to measure or count.)

Lesson Summary

A statistical investigation involves a four-step investigative process:

- Pose questions that can be answered by data.
- Design a plan for collecting appropriate data, and then use the plan to collect data.
- Analyze the data.
- Interpret results and draw valid conclusions from the data to answer the question posed.

Name _____   Date _____

1. What is a statistical question?

2. What are the four steps in a statistical investigation?

Your teacher will outline steps you are expected to complete in the next several days to develop this project. Keep in mind that the first step is to formulate a statistical question. With one of the statistical questions posed in this lesson or with a new one developed in this lesson, describe your question and plan to collect and summarize data. Complete the process as outlined by your teacher.

Example 1: Summary Information from Graphs

Here is a data set of the ages (in years) of 43 participants who ran in a 5-kilometer race.

20	30	30	35	36	34	38	46
45	18	43	23	47	27	21	30
32	32	31	32	36	74	41	41
51	61	50	34	34	34	35	28
57	26	29	49	41	36	37	41
38	30	30					

Here are some summary statistics, a dot plot, and a histogram for the data:

Minimum = 18, Q1 = 30, Median = 35, Q3 = 41, Maximum = 74; Mean = 36.8, MAD = 8.1

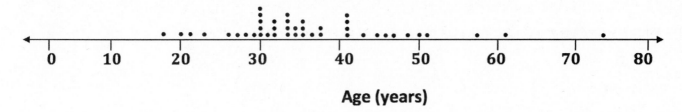

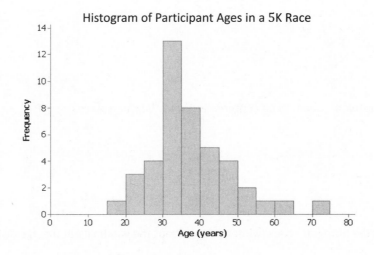

Exercises 1–7

1. Based on the histogram, would you describe the shape of the data distribution as approximately symmetric or as skewed? Would you have reached this same conclusion looking at the dot plot?

2. If there had been 500 participants instead of just 43, would you use a dot plot or a histogram to display the data?

3. What is something you can see in the dot plot that is not as easy to see in the histogram?

4. Do the dot plot and the histogram seem to be centered in about the same place?

5. Do both the dot plot and the histogram convey information about the variability in the age distribution?

Lesson 18:   Connecting Graphical Representations and Numerical
               Summaries

EUREKA
MATH

6. If you did not have the original data set and only had the dot plot and the histogram, would you be able to find the value of the median age from the dot plot?

7. Explain why you would only be able to estimate the value of the median if you only had a histogram of the data.

Exercises 8–12: Graphs and Numerical Summaries

8. Suppose that a newspaper article was written about the race. The article included the histogram shown here and also said, "The race attracted many older runners this year. The median age was 45." Based on the histogram, how can you tell that this is an incorrect statement?

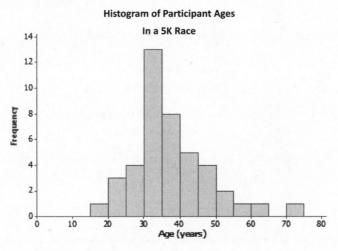

Histogram of Participant Ages In a 5K Race

9. One of the histograms below is another correctly drawn histogram for the runners' ages. Select the correct histogram, and explain how you determined which graph is correct (and which one is incorrect) based on the summary measures and dot plot.

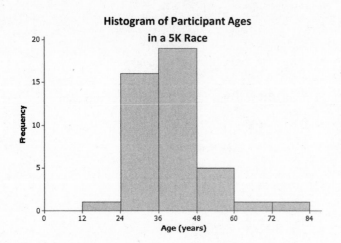

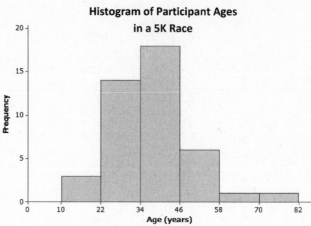

10. The histogram below represents the age distribution of the population of Kenya in 2010.

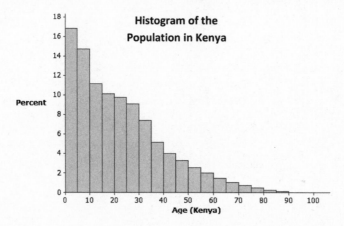

a. How do we know from the graph above that the first quartile (Q1) of this age distribution is between 5 and 10 years of age?

Lesson 18:   Connecting Graphical Representations and Numerical Summaries

EUREKA
MATH

b. Someone believes that the median age in Kenya is about 30. Based on the histogram, is 30 years a good estimate of the median age for Kenya? Explain why it is or why it is not.

11. The histogram below represents the age distribution of the population of the United States in 2010. Based on the histogram, which of the following ranges do you think includes the median age for the United States: 20–30, 30–40, or 40–50? Why?

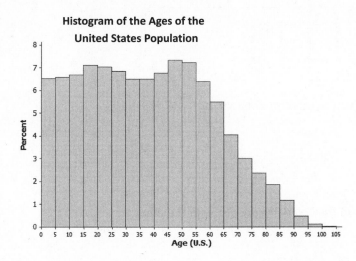

12. Consider the following three dot plots. Note: The same scale is used in each dot plot.

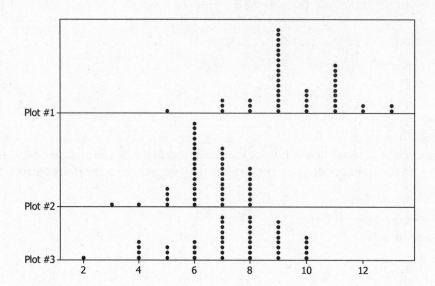

a. Which dot plot has a median of 8? Explain why you selected this dot plot over the other two.

b. Which dot plot has a mean of 9.6? Explain why you selected this dot plot over the other two.

c. Which dot plot has a median of 6 and a range of 5? Explain why you selected this dot plot over the other two.

EUREKA
MATH®

Name _____   Date _____

1. Many states produce maple syrup, which requires tapping sap from a maple tree. However, some states produce more pints of maple syrup per tap than other states. The following dot plot shows the pints of maple syrup yielded per tap in each of the 10 maple syrup–producing states in 2012.

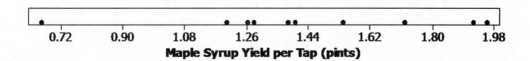

Maple Syrup Yield per Tap by State (10 States - 2012 USDA Summary)

0.72   0.90   1.08   1.26   1.44   1.62   1.80   1.98

Maple Syrup Yield per Tap (pints)

Which *one* of the three sets of summary measures below could be correct summary measures for the data set displayed in the dot plot? For each choice that you eliminate, give at least one reason for eliminating it as a possibility.

a. Minimum = 0.66, Q1 = 1.26, Median = 1.385, Q3 = 1.71, Maximum = 1.95, Range = 2.4; Mean = 1.95, MAD = 0.28

b. Minimum = 0.66, Q1 = 1.26, Median = 1.71, Q3 = 1.92, Maximum = 1.95, Range = 1.29; Mean = 1.43, MAD = 2.27

c. Minimum = 0.66, Q1 = 1.26, Median = 1.385, Q3 = 1.71, Maximum = 1.95, Range = 1.29; Mean = 1.43, MAD = 0.28

2. Which *one* of the three histograms below could be a histogram for the data displayed in the dot plot in Problem 1? For each histogram that you eliminate, give at least one reason for eliminating it as a possibility.

a.

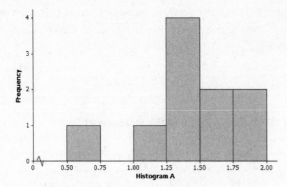

b.

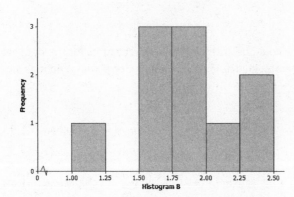

c.

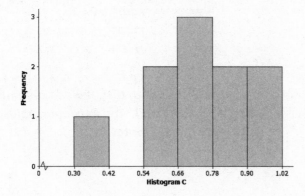

Lesson 18: Connecting Graphical Representations and Numerical Summaries

EUREKA MATH®

1. The following histogram shows the amount of wind produced by Puerto Rico and 39 states that had wind facilities by the end of 2014. Many of these states produced less than 3,000 megawatts of wind, but one state produced over 14,000 megawatts (Texas). For the histogram, which *one* of the three sets of summary measures could match the graph? For each choice that you eliminate, give at least one reason for eliminating the choice.

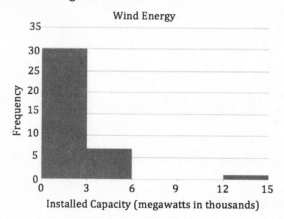

Because a histogram does not show individual values, it is not possible to determine exact values for the 5-number summary. However, I can use my knowledge of what these terms mean to see what makes sense in the context of the situation and the information provided on the histogram.

Data source: http://www.neo.ne.gov/statshtml/205/205_2015.htm, accessed October 29, 2018

a. Minimum $= 1$, $Q1 = 0.16$, Median $= 0.73$, $Q3 = 2.6$, Maximum $= 13$, Mean $= 1.6$, MAD $= 1.5$

b. Minimum $= 1.5$, $Q1 = 16.6$, Median $= 0.73$, $Q3 = 2.6$, Maximum $= 14$, Mean $= 15.2$, MAD $= 1.67$

c. Minimum $= 2.7$, $Q1 = 0.16$, Median $= 730$, $Q3 = 6$, Maximum $= 17$, Mean $= 1.6$, MAD $= 1.5$

The correct answer is (a).

Choice (b) would not work because Q1 (median of the lower half of the data) must be less than 0.73, the median, so a value of 16.6 is unreasonable. Also, the mean cannot be larger than the maximum value listed in the graph, so the value of the mean in choice (b) is not reasonable.

Choice (c) would not work. The value of the median is unreasonable given the scale of the graph. Also, the mean is most likely greater than (not less than) the median given the skewed right nature of the distribution and the large outlier, which is not the case in choice (c). The maximum value is also unreasonable since it is larger than the largest number on the horizontal scale.

2. The heights (rounded to the nearest inch) of the 34 members of the 2015–2016 Brigham Young University Women's Swimming and Diving Team are shown in the dot plot below.

University Women's Swimming and Diving Team

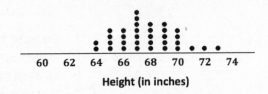

Height (in inches)

Data source: http://byucougars.com/roster/w-swimming-diving, accessed September 28, 2015.

a. Use the dot plot to determine the 5-number summary (minimum, lower quartile, median, upper quartile, and maximum) for the data set.

Min = 64, Q1 = 66, Median = 67.5, Q3 = 69, and Max = 73

> The lower quartile, Q1, is the median of the bottom half of the data.

> The upper quartile, Q3, is the median of the top half of the data.

b. Based on this dot plot, make a histogram of the heights using the following intervals: 64 to <66 inches, 66 to <68 inches, and so on.

Histogram of Heights of BYU's Women's Swimming and Diving Team

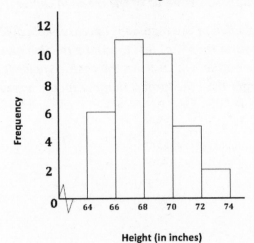

Height (in inches)

EUREKA MATH

1. The following histogram shows the amount of coal produced (by state) for the 20 largest coal-producing states in 2011. Many of these states produced less than 50 million tons of coal, but one state produced over 400 million tons (Wyoming). For the histogram, which *one* of the three sets of summary measures could match the graph? For each choice that you eliminate, give at least one reason for eliminating the choice.

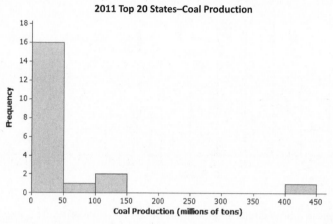

2011 Top 20 States–Coal Production

Source: National Mining Association (2013) from http://www.nma.org/pdf/c_production_state_rank.pdf accessed May 5, 2013

a. Minimum = 1, Q1 = 12, Median = 36, Q3 = 57, Maximum = 410; Mean = 33, MAD = 2.76

b. Minimum = 2, Q1 = 13.5, Median = 27.5, Q3 = 44, Maximum = 439; Mean = 54.6, MAD = 52.36

c. Minimum = 10, Q1 = 37.5, Median = 62, Q3 = 105, Maximum = 439; Mean = 54.6, MAD = 52.36

2. The heights (rounded to the nearest inch) of the 41 members of the 2012–2013 University of Texas Men's Swimming and Diving Team are shown in the dot plot below.

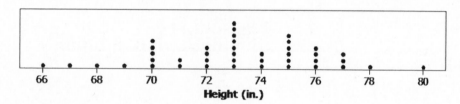

Height (in.)

Source: http://www.texassports.com accessed April 30, 2013

a. Use the dot plot to determine the 5-number summary (minimum, lower quartile, median, upper quartile, and maximum) for the data set.

b. Based on this dot plot, make a histogram of the heights using the following intervals: 66 to <68 inches, 68 to <70 inches, and so on.

3. Data on the weight (in pounds) of 143 wild bears are summarized in the histogram below.

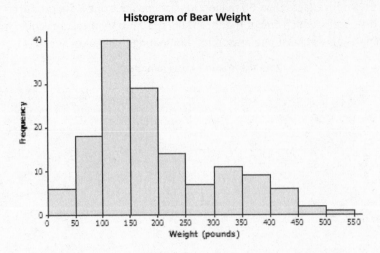

Which *one* of the three dot plots below could be a dot plot of the bear weight data? Explain how you determined which the correct plot is.

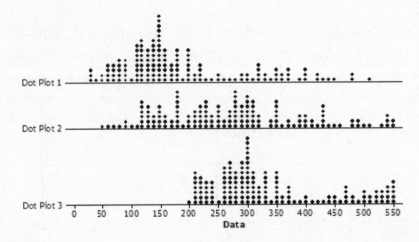

Suppose that you are interested in comparing the weights of adult male polar bears and the weights of adult male grizzly bears. If data were available on the weights of these two types of bears, they could be used to answer questions such as:

Do adult polar bears typically weigh less than adult grizzly bears?

Are the weights of adult polar bears similar to each other, or do the weights tend to differ a lot from bear to bear?

Are the weights of adult polar bears more consistent than the weights of adult grizzly bears?

These questions could be answered most easily by comparing the weight distributions for the two types of bears. Graphs of the data distributions (such as dot plots, box plots, or histograms) that are drawn side by side and that are drawn to the same scale make it easy to compare data distributions in terms of center, variability, and shape.

In this lesson, when two or more data distributions are presented, think about the following:

How are the data distributions similar?

How are the data distributions different?

What do the similarities and differences tell you in the context of the data?

Example 1: Comparing Groups Using Box Plots

Recall that a *box plot* is a visual representation of a five-number summary. The box part of a box plot is drawn so that the width of the box represents the IQR. The distance from the far end of the line on the left to the far end of the line on the right represents the range.

If two box plots (each representing a different distribution) are drawn side by side using the same scale, it is easy to compare the values in the five-number summaries for the two distributions and to visually compare the IQRs and ranges.

Here is a data set of the ages of 43 participants in a 5-kilometer race (shown in a previous lesson).

20	30	30	35	36	34	38	46
45	18	43	23	47	27	21	30
32	32	31	32	36	74	41	41
51	61	50	34	34	34	35	28
57	26	29	49	41	36	37	41
38	30	30					

Here is the five-number summary for the data: Minimum = 18, Q1 = 30, Median = 35, Q3 = 41, Maximum = 74.

There was also a 15-kilometer race. The ages of the 55 participants in that race appear below.

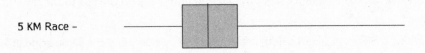

47	19	30	30	36	37	35	39
19	49	47	16	45	22	50	27
19	20	30	32	32	31	32	37
22	81	43	43	54	66	53	35
22	35	35	36	28	61	26	29
38	52	43	37	38	43	39	30
58	30	48	49	54	56	58	

Does the longer race appear to attract different runners in terms of age? Below are side-by-side box plots that may help answer that question. Side-by-side box plots are two or more box plots drawn using the same scale. What do you notice about the two box plots?

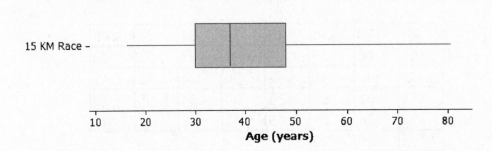

Lesson 19:   Comparing Data Distributions

© 2019 Great Minds®. eureka-math.org

EUREKA MATH®

Exercises 1–6

1. Based on the box plots, estimate the values in the five-number summary for the age in the 15-kilometer race data set.

2. Do the two data sets have the same median? If not, which race had the higher median age?

3. Do the two data sets have the same IQR? If not, which distribution has the greater spread in the middle 50% of its distribution?

4. Which race had the smaller overall range of ages? What do you think the range of ages is for the 15-kilometer race?

5. Which race had the oldest runner? About how old was this runner?

6. Now, consider just the youngest 25% of the runners in the 15-kilometer race. How old was the youngest runner in this group? How old was the oldest runner in this group? How does that compare with the 5-kilometer race?

Exercises 7–12: Comparing Box Plots

In 2012, Major League Baseball had two leagues: an American League of 14 teams and a National League of 16 teams. Jesse wondered if American League teams have higher batting averages and on-base percentages. (Higher values are better.) Use the following box plots to investigate. (Source: http://mlb.mlb.com/stats/sortable.jsp, accessed May 13, 2013)

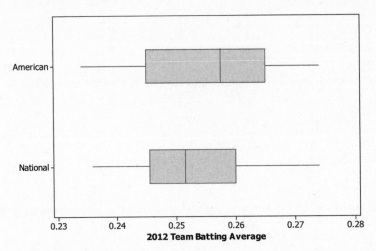

7. Was the highest American League team batting average very different from the highest National League team batting average? Approximately how large was the difference, and which league had the higher maximum value?

8. Was the range of American League team batting averages very different or only slightly different from the range of National League team batting averages?

9. Which league had the higher median team batting average? Given the scale of the graph and the range of the data sets, does the difference between the median values for the two leagues seem to be small or large? Explain why you think it is small or large.

EUREKA
MATH

10. Based on the box plots below for on-base percentage, which three summary values (from the five-number summary) appear to be the same or virtually the same for both leagues?

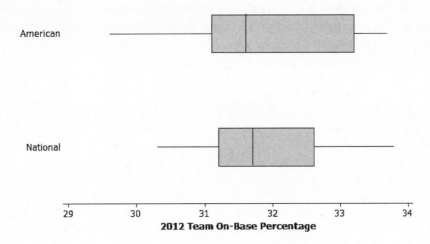

11. Which league's data set appears to have less variability? Explain.

12. Recall that Jesse wondered if American League teams have higher batting averages and on-base percentages. Based on the box plots given above, what would you tell Jesse?

Lesson Summary

When comparing the distribution of a quantitative variable for two or more distinct groups, it is useful to display the groups' distributions side by side using graphs drawn to the same scale. This makes it easier to describe the similarities and differences in the distributions of the groups.

Name _____   Date _____

Hay is used to feed animals such as cows, horses, and goats. Almost $\frac{1}{3}$ of the hay grown in the United States comes from just five states. Is this because these states have more acres planted in hay, or could it be because these states produce more hay per acre than other states? The following box plots show the distribution of hay produced (in tons) per acre planted in hay for three different regions: 22 eastern states, 14 midwestern states, and 12 western states.

Source: *United States Department of Agriculture National Agricultural Statistics Service Crop Production 2012 Summary*, ISSN: 1057-7823, p. 75, accessed May 5, 2013

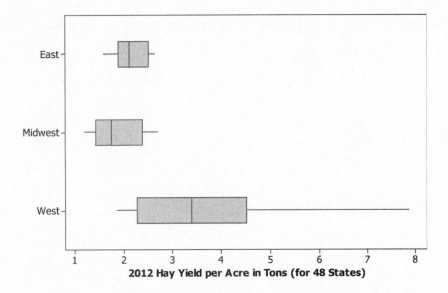

1. Which of the three regions' data sets has the least variability? Which has the greatest variability? To explain how you chose your answers, write a sentence or two that supports your choices by comparing relevant summary measures (such as median and IQR) or aspects of the graphical displays (such as shape and variability).

2. True or false: The western state with the smallest hay yield per acre has a higher hay yield per acre than at least half of the midwestern states. Explain how you know this is true or how you know this is false.

3. Which region typically has states with the largest hay yield per acre? To explain how you chose your answer, write a sentence or two that supports your choice by comparing relevant summary measures or aspects of the graphical displays.

College athletic programs are separated into divisions based on school size, available athletic scholarships, and other factors. A researcher wondered if members of Division I women's basketball programs (usually large schools that offer athletic scholarships) tend to be taller than members of Division III women's basketball programs (usually smaller schools that do not offer athletic scholarships). To begin the investigation, the researcher creates side-by-side box plots for the heights (in inches) of 35 female basketball players at Blue Ridge University (a Division I program) and the heights (in inches) of 15 female basketball players at Stanley Mountain College (a Division III program).

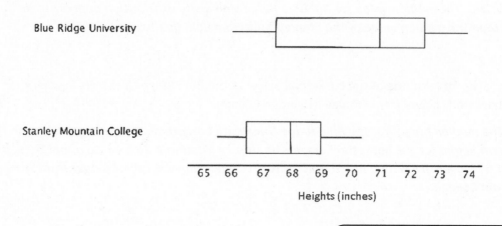

a. Which data set has the smaller range?

    Stanley Mountain College has the smaller range.

    > From the minimum value to the maximum value, the range is smaller for Stanley Mountain College than Blue Ridge University.

b. True or false: A basketball player who had a height equal to the median for the Blue Ridge University would be taller than the median height of basketball players at Stanley Mountain College.

    True

    > The median height for a player from Blue Ridge University is 71 inches, and the median height for a player from Stanley Mountain College is 68 inches.

c. To be thorough, the researcher will examine many other colleges' sports programs to further investigate the claim that members of Division I women's basketball programs are generally taller than the members of Division III women's basketball programs. But given the graph above, in this initial stage of her research, do you think that the claim might be valid? Carefully support your answer using summary measures or graphical attributes.

Based on just these two teams, it looks like the claim may be correct. A large portion of the Blue Ridge University distribution is higher than the maximum value of the Stanley Mountain College distribution. The median value for the Blue Ridge University distribution appears to be 3 inches higher than the median value of the Stanley Mountain College distribution.

d. True or False: At least one of the basketball players from Blue Ridge University is taller than the tallest basketball player from Stanley Mountain College.

True. The median height for the Blue Ridge University basketball players is greater than the maximum height for the basketball players at Stanley Mountain College, so about 50% of the basketball players from Blue Ridge University are taller than the tallest player from Stanley Mountain College.

EUREKA
MATH

1. College athletic programs are separated into divisions based on school size, available athletic scholarships, and other factors. A researcher wondered if members of swimming and diving programs in Division I (usually large schools that offer athletic scholarships) tend to be taller than the swimmers and divers in Division III programs (usually smaller schools that do not offer athletic scholarships). To begin the investigation, the researcher creates side-by-side box plots for the heights (in inches) of 41 male swimmers and divers at Mountain Vista University (a Division I program) and the heights (in inches) of 10 male swimmers and divers at Eaglecrest College (a Division III program).

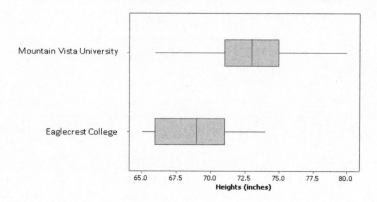

a. Which data set has the smaller range?

b. True or false: A swimmer who had a height equal to the median for the Mountain Vista University would be tallerthan the median height of swimmers and divers at Eaglecrest College.

c. To be thorough, the researcher will examine many other colleges' sports programs to further investigate the claim that members of swimming and diving programs in Division I are generally taller than the swimmers and divers in Division III. But given the graph above, in this initial stage of her research, do you think that the claim might be valid? Carefully support your answer using summary measures or graphical attributes.

2. Data on the weights (in pounds) of 100 polar bears and 50 grizzly bears are summarized in the box plots shown below.

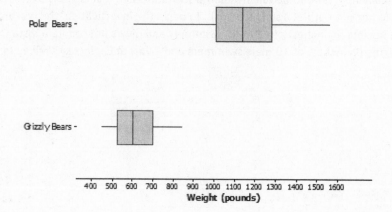

a. True or false: At least one of the polar bears weighed more than the heaviest grizzly bear. Explain how you know.

b. True or false: Weight differs more from bear to bear for polar bears than for grizzly bears. Explain how you know.

c. Which type of bear tends to weigh more? Explain.

Lesson 19:   Comparing Data Distributions

3. Many movie studios rely heavily on viewer data to determine how a movie will be marketed and distributed. Recently, previews of a soon-to-be-released movie were shown to 300 people. Each person was asked to rate the movie on a scale of 0 to 10, with 10 representing "best movie I have ever seen" and 0 representing "worst movie I have ever seen."

    Below are some side-by-side box plots that summarize the ratings by gender and by age.

    For 150 women and 150 men:

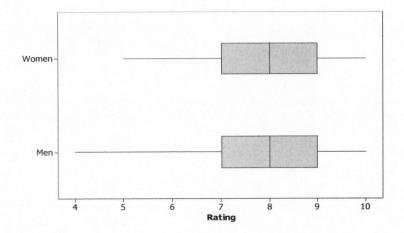

    For 3 age groups:

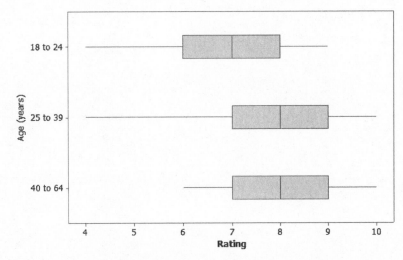

    a. Does it appear that the men and women rated the film in a similar manner or in a very different manner? Write a few sentences explaining your answer using comparative information about center and variability.

    b. It appears that the film tended to receive better ratings from the older members of the group. Write a few sentences using comparative measures of center and spread or aspects of the graphical displays to justify this claim.

Great Lakes yellow perch are fish that live in each of the five Great Lakes and many other lakes in the eastern and upper Great Lakes regions of the United States and Canada. Both countries are actively involved in efforts to maintain a healthy population of perch in these lakes.

Example 1: The Great Lakes Yellow Perch

Scientists collected data from many yellow perch because they were concerned about the survival of the yellow perch. What data do you think researchers might want to collect about these perch?

Scientists captured yellow perch from a lake in this region. They recorded data on each fish and then returned each fish to the lake. Consider the following histogram of data on the length (in centimeters) for a sample of yellow perch.

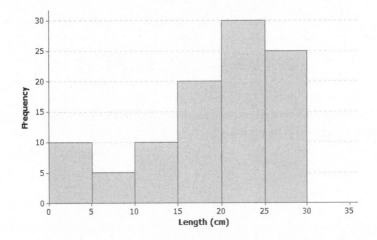

Exercises 1–11

Scientists were concerned about the survival of the yellow perch as they studied the histogram.

1. What statistical question could be answered based on this data distribution?

2. Use the histogram to complete the following table:

Length of Fish in Centimeters (cm)	Number of Fish
$0 \leq 5$ cm	
$5 \leq 10$ cm	
$10 \leq 15$ cm	
$15 \leq 20$ cm	
$20 \leq 25$ cm	
$25 \leq 30$ cm	

3. The length of each fish in the sample was measured and recorded before the fish was released back into the lake. How many yellow perch were measured in this sample?

4. Would you describe the distribution of the lengths of the fish in the sample as a skewed distribution or as an approximately symmetric distribution? Explain your answer.

EUREKA MATH

5. What percentage of fish in the sample were less than 10 centimeters in length?

6. If the smallest fish in this sample was 2 centimeters in length, what is your estimate of an interval of lengths that would contain the lengths of the shortest 25% of the fish? Explain how you determined your answer.

7. If the length of the largest yellow perch was 29 centimeters, what is your estimate of an interval of lengths that would contain the lengths of the longest 25% of the fish?

8. Estimate the median length of the yellow perch in the sample. Explain how you determined your estimate.

9. Based on the shape of this data distribution, do you think the mean length of a yellow perch would be greater than, less than, or the same as your estimate of the median? Explain your answer.

10. Recall that the mean length is the balance point of the distribution of lengths. Estimate the mean length for this sample of yellow perch.

11. The length of a yellow perch is used to estimate the age of the fish. Yellow perch typically grow throughout their lives. Adult yellow perch have lengths between 10 and 30 centimeters. How many of the yellow perch in this sample would be considered adult yellow perch? What percentage of the fish in the sample are adult fish?

Example 2: What Would a Better Distribution Look Like?

Yellow perch are part of the food supply of larger fish and other wildlife in the Great Lakes region. Why do you think that the scientists worried when they saw the histogram of fish lengths given previously in Exercise 2.

Sketch a histogram representing a sample of 100 yellow perch lengths that you think would indicate the perch are not in danger of dying out.

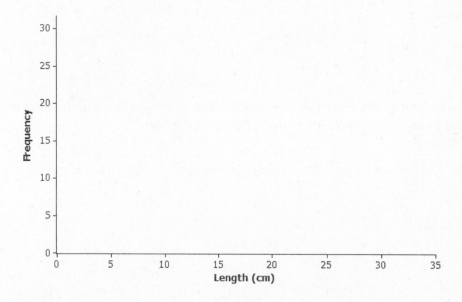

Lesson 20:   Describing Center, Variability, and Shape of a Data
Distribution from a Graphical Representation

© 2019 Great Minds®. eureka-math.org

Exercises 12–17: Estimating the Variability in Yellow Perch Lengths

You estimated the median length of yellow perch from the first sample in Exercise 8. It is also useful to describe variability in the length of yellow perch. Why might this be important? Consider the following questions:

12. In several previous lessons, you described a data distribution using the five-number summary. Use the histogram and your answers to the questions in previous exercises to provide estimates of the values for the five-number summary for this sample:

    Minimum (min) value =

    Q1 value =

    Median =

    Q3 value =

    Maximum (max) value =

13. Based on the five-number summary, what is an estimate of the value of the interquartile range (IQR) for this data distribution?

14. Sketch a box plot representing the lengths of the yellow perch in this sample.

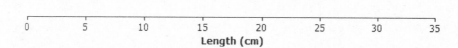

EUREKA
MATH®

Lesson 20:   Describing Center, Variability, and Shape of a Data
               Distribution from a Graphical Representation

255

© 2019 Great Minds®. eureka-math.org

15. Which measure of center, the median or the mean, is closer to where the lengths of yellow perch tend to cluster?

16. What value would you report as a typical length for the yellow perch in this sample?

17. The mean absolute deviation (or MAD) or the interquartile range (IQR) is used to describe the variability in a data distribution. Which measure of variability would you use for this sample of perch? Explain your answer.

EUREKA
MATH

Lesson Summary

Data distributions are usually described in terms of shape, center, and spread. Graphical displays such as histograms, dot plots, and box plots are used to assess the shape. Depending on the shape of a data distribution, different measures of center and variability are used to describe the distribution. For a distribution that is skewed, the median is used to describe a typical value, whereas the mean is used for distributions that are approximately symmetric. The IQR is used to describe variability for a skewed data distribution, while the MAD is used to describe variability for a distribution that is approximately symmetric.

Name _____   Date _____

1. Great Lake yellow perch continue to grow until they die. What does the histogram in Example 1 indicate about the ages of the perch in the sample?

2. What feature of the histogram in Example 1 indicates that the values of the mean and the median of the data distribution will not be equal?

3. Adult yellow perch have lengths between 10 and 30 centimeters. Would a perch with a length equal to the median length be classified as an adult or a pre-adult fish? Explain your answer.

Data was collected on the heights of black bears in a particular forest. A histogram of the heights for the black bear in this sample is shown below.

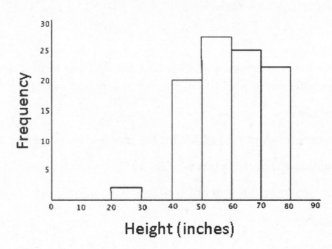

Heights of Black Bears

1. If the height of an average adult black bear is 45 to 75 inches, what can you conclude about this sample?
    Explain your answer.

    Answers may vary. One response might be: In this sample, there were most likely **2** *to* **3** *baby black bears that were measured since* **2** *to* **3** *bears were between* **20** *and* **30** *inches tall and much shorter than the average height.*

2. Does this histogram represent a data distribution that is skewed or that is nearly symmetrical?

    This distribution is skewed. The tail of this distribution is to the left, or toward the shorter heights.

    > In a skewed shape, there are values that are unusual (or not typical) when compared to the rest of the data. In this histogram, there are values much lower than most of the data.

3. What measure of center would you use to describe a typical height of a black bear in this sample? Explain your answer.

   I recommend the median of the data distribution as a description of a typical value of the height of a black bear because this distribution is skewed.

4. Assume the smallest black bear measured was 24 inches tall, and the largest black bear measured was 78 inches tall. Estimate the values in the five-number summary for this sample:

   Minimum (min) = 24 inches

   Q1 = 45 inches (a value greater than 40 but within the interval of 40 to less than 50 inches)

   Median = 60 inches (a value within the interval of 50 to 70 inches)

   Q3 = 70 inches (a value within the interval of 70 to less than 80 inches)

   Maximum (max) = 78 inches

   > Since histograms do not show specific values, it is hard to determine values for the five-number summary. The minimum and maximum are given so these are specific values but the values for Q1, median, and Q3 are reasonable estimates.

5. Based on the shape of this data distribution, do you think the mean height of a black bear from this sample would be greater than, less than, or the same as your estimate of the median? Explain your answer.

   An estimate of the mean would be less than the median height because the values in the tail, or to the left of the median, pull the mean in that direction.

   > When there are uncharacteristically small or large values in a data set, the mean is sensitive to these values, and the mean may be pulled toward the very small or large values.

Lesson 20:   Describing Center, Variability, and Shape of a Data
                        Distribution from a Graphical Representation

EUREKA MATH

6. Estimate the mean value of this data distribution.

   An estimate of the mean would be a value slightly smaller than the median value. For example, a mean of 55 inches would be a reasonable estimate of a balance point.

   > Since I estimated the median to be 60 inches and I know the mean will be slightly less because of the tail toward the shorter heights (the left), I can choose a value of 55 inches to represent the mean.

7. What is your estimate of a typical height of a black bear in this sample? Did you use the mean height or median height for this estimate? Explain.

   Since the median was selected as the appropriate estimate of a measure of center because this data distribution is skewed, a value of 60 inches (or whatever students used to estimate the median) would be an estimate of a typical height for a black bear from this sample.

   > The measure of center (mean or median) represents a typical value for a data set.

8. Would you use the MAD or the IQR to describe variability in the height of black bears in this sample? Estimate the value of the measure of variability that you selected.

   I would use the IQR to describe the variability because the data distribution is skewed, and the median was used as a typical height for a black bear. An estimate of the IQR based on the above estimates would be as follows: 70 inches − 45 inches = 25 inches.

   > $IQR = Q3 - Q1$

Another sample of Great Lake yellow perch from a different lake was collected. A histogram of the lengths for the fish in this sample is shown below.

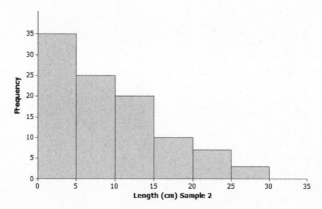

1. If the length of a yellow perch is an indicator of its age, how does this second sample differ from the sample you investigated in the exercises? Explain your answer.

2. Does this histogram represent a data distribution that is skewed or that is nearly symmetrical?

3. What measure of center would you use to describe a typical length of a yellow perch in this second sample? Explain your answer.

4. Assume the smallest perch caught was 2 centimeters in length, and the largest perch caught was 29 centimeters in length. Estimate the values in the five-number summary for this sample:

    Minimum (min) value =

    Q1 value =

    Median value =

    Q3 value =

    Maximum (max) value =

5. Based on the shape of this data distribution, do you think the mean length of a yellow perch from this second sample would be greater than, less than, or the same as your estimate of the median? Explain your answer.

6. Estimate the mean value of this data distribution.

7. What is your estimate of a typical length of a yellow perch in this sample? Did you use the mean length from Problem 5 for this estimate? Explain why or why not.

8. Would you use the MAD or the IQR to describe variability in the length of Great Lakes yellow perch in this sample? Estimate the value of the measure of variability that you selected.

Each of the lessons in this module is about data. What are data? What questions can be answered by data? How do you represent the data distribution so that you can understand and describe its shape? What does the shape tell us about how to summarize the data? What is a typical value of the data set? These and many other questions were part of your work in the exercises and investigations. There is still a lot to learn about what data tell us. You will continue to work with statistics and probability in Grades 7 and 8 and throughout high school, but you have already begun to see how to uncover the stories behind data.

When you started this module, the four steps used to carry out a statistical study were introduced.

   Step 1: Pose a question that can be answered by data.

   Step 2: Collect appropriate data.

   Step 3: Summarize the data with graphs and numerical summaries.

   Step 4: Answer the question posed in Step 1 using the numerical summaries and graphs.

In this lesson, you will carry out these steps using a given data set.

Exploratory Challenge: Annual Rainfall in the State of New York

The National Climate Data Center collects data throughout the United States that can be used to summarize the climate of a region. You can obtain climate data for a state, a city, a county, or a region. If you were interested in researching the climate in your area, what data would you collect? Explain why you think these data would be important in a statistical study of the climate in your area.

For this lesson, you will use yearly rainfall data for the state of New York that were compiled by the National Climate Data Center. The following data are the number of inches of rain (averaged over various locations in the state) for the years from 1983 to 2012 (30 years).

45	42	39	44	39	35	42	49	37	42	41	42	37	50	39
41	38	46	34	44	48	50	47	49	44	49	43	44	54	40

Use the four steps to carry out a statistical study using these data.

Step 1: Pose a question that can be answered by data.

What is a statistical question that you think can be answered with these data? Write your question in the template provided for this lesson.

Step 2: Collect appropriate data.

The data have already been collected for this lesson. How do you think these data were collected? Recall that the data are the number of inches of rain (averaged over various locations in the state) for the years from 1983 to 2012 (30 years). Write a summary of how you think the data were collected in the template for this lesson.

Step 3: Summarize the data with graphs and numerical summaries.

A good first step might be to summarize the data with a dot plot. What other graph might you construct? Construct a dot plot or another appropriate graph in the template for this lesson.

What numerical summaries will you calculate? What measure of center will you use to describe a typical value for these data? What measure of variability will you calculate and use to summarize the variability of the data? Calculate the numerical summaries, and write them in the template for this lesson.

Step 4: Answer your statistical question using the numerical summaries and graphs.

Write a summary that answers the question you posed in the template for this lesson.

Lesson 21:   Summarizing a Data Distribution by Describing Center, Variability, and Shape

EUREKA MATH

Template for Lesson 21

Step 1: Pose a question that can be answered by data.

Step 2: Collect appropriate data.

Step 3: Summarize the data with graphs and numerical summaries.

Construct at least one graph of the data distribution. Calculate appropriate numerical summaries of the data. Also, indicate why you selected these summaries.

Step 4: Answer your statistical question using the numerical summaries and graphs.

Lesson 21:   Summarizing a Data Distribution by Describing Center, Variability, and Shape

269

Additional Resource Materials

The following could be used to provide structure in constructing a dot plot, histogram, or box plot of the rainfall data. A similar type of grid (or graph paper) could be prepared for students as they complete the Problem Set. The grid provided for students should not include the units along the horizontal axis since that is part of what they are expected to do in preparing their summaries.

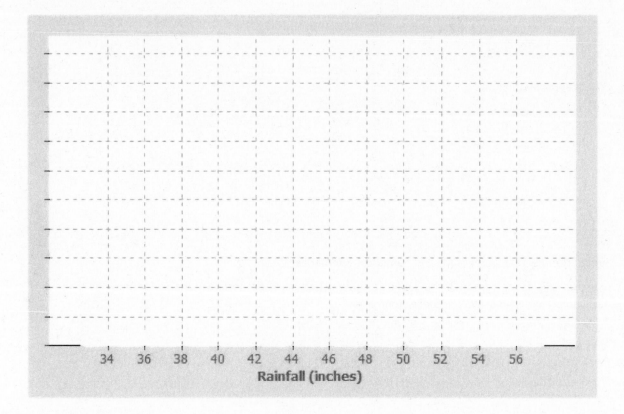

Lesson 21:   Summarizing a Data Distribution by Describing Center,
              Variability, and Shape

EUREKA
MATH®

The following table could be used for students requiring some structure in calculating the mean absolute deviation, or MAD, for the rainfall data.

Data Value	Distance from the Mean
45	
42	
39	
44	
39	
35	
42	
49	
37	
42	
41	
42	
37	
50	
39	
41	
38	
46	
34	
44	
48	
50	
47	
49	
44	
49	
43	
44	
54	
40	

Lesson Summary

Statistics is about using data to answer questions. The four steps used to carry out a statistical study include posing a question that can be answered by data, collecting appropriate data, summarizing the data with graphs and numerical summaries, and using the data, graphs, and summaries to answer the statistical question.

Lesson 21:   Summarizing a Data Distribution by Describing Center, Variability, and Shape

EUREKA
MATH

Name _____   Date _____

Based on the statistical question you are investigating for your project, summarize the four steps you are expected to complete as part of the presentation of your statistical study.

In Lesson 17, you posed a statistical question and created a plan to collect data to answer your question. You also constructed graphs and calculated numerical summaries of your data. Review the data collected and your summaries.

Based on directions from your teacher, create a poster or an outline for a presentation using your own data. On your poster, indicate your statistical question. Also, indicate a brief summary of how you collected your data based on the plan you proposed in Lesson 17. Include a graph that shows the shape of the data distribution, along with summary measures of center and variability. Finally, answer your statistical question based on the graphs and the numerical summaries.

Share the poster you will present in Lesson 22 with your teacher. If you are instructed to prepare an outline of the presentation, share your outline with your teacher.

A statistical study involves the following four-step investigative process:

Step 1:   Pose a question that can be answered by data.

Step 2:   Collect appropriate data.

Step 3:   Summarize the data with graphs and numerical summaries.

Step 4:   Answer the question posed in Step 1 using the numerical summaries and graphs.

Now it is your turn to be a researcher and to present your own statistical study. In Lesson 17, you posed a statistical question, proposed a plan to collect data to answer the question, and collected the data. In Lesson 21, you created a poster or an outline of a presentation that included the following: the statistical question, the plan you used to collect the data, graphs and numerical summaries of the data, and an answer to the statistical question based on your data. Use the following table to organize your presentation.

Points to Consider:	Notes to Include in Your Presentation:
(1) Describe your statistical question.	
(2) Explain to your audience why you were interested in this question.	
(3) Explain the plan you used to collect the data.	
(4) Explain how you organized the data you collected.	

(5)	Explain the graphs you prepared for your presentation and why you made these graphs.	
(6)	Explain what measure of center and what measure of variability you selected to summarize your study. Explain why you selected these measures.	
(7)	Describe what you learned from the data. (Be sure to include an answer to the question from Step (1) on the previous page.)	

Lesson Summary

Statistics is about using data to answer questions. The four steps used to carry out a statistical study include posing a question that can be answered by data, collecting appropriate data, summarizing the data with graphs and numerical summaries, and using the data, graphs, and numerical summaries to answer the statistical question.

Step 1:   What was the statistical question presented on this poster?

Step 2:   How were the data collected?

Step 3:   What graphs and numerical summaries were used to summarize data?

Describe at least one graph presented on the poster. (For example, was it a dot plot? What was represented on the scale?) What numerical summaries of the data were included (e.g., the mean or the median)? Also, indicate why these particular numerical summaries were selected.

Step 4:   Summarize the answer to the statistical question.

Name _____   Date _____

After you have presented your study, consider what your next steps are by answering the following questions:

1. What questions still remain after you concluded your statistical study?

2. What statistical question would you like to answer next as a follow-up to this study?

3. How would you collect the data to answer the new question you posed in Question 2?

Credits

Great Minds® has made every effort to obtain permission for the reprinting of all copyrighted material. If any owner of copyrighted material is not acknowledged herein, please contact Great Minds for proper acknowledgment in all future editions and reprints of this module.